Molecular Docking: A Formula Handbook

N.B. Singh

DEDICATION

To Nature,

I dedicate this book to you, the source of all life. You are my inspiration, my teacher, and my friend.

Thank you for teaching me about the beauty of the world around me. Thank you for showing me the power of the natural world. Thank you for giving me a sense of peace and tranquillity.

I promise to do my part to protect you and your many wonders. I will teach my children about the importance of conservation and sustainability. I will work to make the world a better place for all living things.

Thank you for everything, Nature.

With love,

N.B Singh

Contents

1 **Introduction**   1

   1.1 Background 1

   1.2 Scope and Objectives 2

       1.2.1 Understanding Molecular Interactions 3

       1.2.2 Computational Methods and Algorithms 3

       1.2.3 Data Preparation and Validation Techniques 3

       1.2.4 Applications in Drug Discovery 3

       1.2.5 Integration with Bioinformatics Tools 5

       1.2.6 Challenges and Future Directions 6

2 **Basics of Molecular Docking**   9

   2.1 Molecular Docking Overview 9

       2.1.1 Binding Affinity and Energetics 9

       2.1.2 Docking Algorithm Types 10

       2.1.3 Scoring Functions 12

       2.1.4 Ligand and Receptor Preparation 14

       2.1.5 Visualization of Docking Results 15

       2.1.6 Challenges and Limitations 16

       2.1.7 Applications in Drug Discovery 16

       2.1.8 Case Studies 18

       2.1.9 Emerging Trends 19

   2.2 Key Concepts 19

2.2.1   Binding Affinity   20

2.2.2   Scoring Functions   20

2.2.3   Docking Algorithm   20

2.2.4   Ligand and Receptor Representations   22

3   Computational Methods   **25**

3.1   Force Fields   25

3.1.1   The Foundation of Force Fields   25

3.1.2   Bonded Interactions: Springs and Angles   27

3.1.3   Non-Bonded Interactions: Van der Waals and Electro-
          static Forces   28

3.1.4   Combining Forces: The Total Potential Energy   30

3.1.5   Popular Force Fields in Molecular Docking   31

3.1.6   Parameterization of Force Fields   31

3.1.7   Validation and Benchmarking   32

3.1.8   Limitations of Force Fields   34

3.1.9   Force Fields in Molecular Docking Simulations   35

3.1.10   Advanced Force Field Techniques   37

3.1.11   Case Studies: Force Field Applications   38

3.1.12   Machine Learning and Force Fields   39

3.1.13   Future Directions in Force Field Development   40

3.2   Search Algorithms   41

3.2.1   Importance of Search Algorithms   41

3.2.2   Exhaustive Search Methods   42

3.2.3   Monte Carlo Methods   43

3.2.4   Genetic Algorithms   44

3.2.5   Lamarckian Genetic Algorithms   45

3.2.6   Ant Colony Optimization   46

3.2.7   Hybrid Search Strategies   47

3.2.8   Adaptive Search Algorithms   47

3.2.9   Parallelization and High-Performance Computing   48

3.2.10 Case Studies: Search Algorithm Applications 49

3.2.11 Machine Learning in Search Algorithms 50

3.2.12 Validation and Benchmarking 51

3.2.13 Limitations of Search Algorithms 53

3.2.14 Future Directions in Search Algorithm Development . . . 54

4 Data Preparation 57

4.1 Preparation of Ligands 57

4.1.1 Structure Retrieval 57

4.1.2 Geometry Optimization 58

4.1.3 Tautomer and Stereoisomer Handling 59

4.1.4 Ionization State Determination 60

4.1.5 Parameterization for Molecular Mechanics 61

4.1.6 Adding Hydrogen Atoms 62

4.1.7 Flexible Ligand Considerations 63

4.1.8 Protonation State and Tautomer Sensitivity 64

4.1.9 Ligand Covalent Modifications 64

4.1.10 Charge Neutralization 65

4.1.11 Validation of Prepared Ligands 66

4.1.12 Case Studies: Ligand Preparation in Drug Discovery . . . 67

4.1.13 Machine Learning Applications in Ligand Preparation . . 67

4.1.14 Challenges and Considerations in Ligand Preparation . . 68

4.1.15 Integration with Receptor Preparation 69

4.1.16 Future Directions in Ligand Preparation 69

4.2 Preparation of Receptors 70

4.2.1 Protein Retrieval and Structure Selection 70

4.2.2 Removal of Water Molecules and Heteroatoms 71

4.2.3 Addition of Hydrogen Atoms 71

4.2.4 Protonation State Determination 72

4.2.5 Energy Minimization 72

4.2.6 Handling of Missing Side Chains 72

4.2.7   Ligand Binding Site Identification 72

4.2.8   Flexible Residue Considerations 72

4.2.9   Co-factor and Metal Ion Handling 73

4.2.10 Validation of Prepared Receptors 73

4.2.11 Case Studies: Receptor Preparation in Drug Discovery . . 73

4.2.12 Machine Learning Applications in Receptor Preparation . 73

4.2.13 Challenges and Considerations in Receptor Preparation . 74

4.2.14 Integration with Ligand Preparation 74

4.2.15 Future Directions in Receptor Preparation 74

5 Validation and Evaluation   75

5.1   Validation Metrics 75

5.1.1   Root Mean Square Deviation (RMSD) 75

5.1.2   Ligand Efficiency 76

5.1.3   Area Under the Receiver Operating Characteristic (AU-ROC) 76

5.1.4   Enrichment Factor (EF) 76

5.1.5   Consensus Scoring 76

5.1.6   Specificity and Sensitivity 77

5.1.7   Kendall's Tau 77

5.1.8   Receiver Operating Characteristic (ROC) Curve 77

5.1.9   Predictive Index (PI) 78

5.1.10 Success Rate 78

5.1.11 F-Measure 78

5.1.12 Matthews Correlation Coefficient (MCC) 78

5.1.13 Distance-Dependent Metrics 79

5.1.14 Case Studies: Application of Validation Metrics 79

5.1.15 Machine Learning for Metric Optimization 79

5.1.16 Limitations and Considerations in Metric Selection 79

5.1.17 Integration with Experimental Data 79

5.1.18 Future Directions in Validation Metrics 80

5.2 Benchmark Datasets 80

    5.2.1 Purpose of Benchmark Datasets 80

    5.2.2 Characteristics of Ideal Benchmark Datasets 80

    5.2.3 Commonly Used Benchmark Datasets 81

    5.2.4 Metrics for Benchmark Evaluation 81

    5.2.5 Case Studies: Benchmark Dataset Applications 82

    5.2.6 Creation of Customized Benchmark Datasets 82

    5.2.7 Machine Learning in Benchmark Dataset Selection 82

    5.2.8 Limitations of Benchmark Datasets 83

    5.2.9 Integration with Experimental Data 83

    5.2.10 Future Directions in Benchmarking 83

    5.2.11 Community Involvement in Benchmarking 83

    5.2.12 Ensuring Reproducibility in Benchmarking 83

6 Advanced Topics   **85**

6.1 Protein-Ligand Interactions 85

    6.1.1 Forces Governing Protein-Ligand Interactions 85

    6.1.2 Quantifying Binding Affinities 86

    6.1.3 Key Binding Motifs 87

    6.1.4 Computational Methods for Analysis 87

    6.1.5 Example: Hydrogen Bonding in Drug Design 88

    6.1.6 Challenges in Predicting Interactions 88

    6.1.7 Future Directions in Protein-Ligand Interactions 88

6.2 Flexible Docking 88

    6.2.1 Importance of Flexibility in Molecular Docking 89

    6.2.2 Mathematical Modeling of Flexible Docking 89

    6.2.3 Scoring Functions in Flexible Docking 89

    6.2.4 Handling Protein Flexibility 90

    6.2.5 Example: Induced Fit Docking 90

    6.2.6 Incorporating Ligand Flexibility 90

    6.2.7 Mathematical Formulation of Ligand Flexibility 90

6.2.8   Hybrid Approaches in Flexible Docking   91

6.2.9   Challenges in Flexible Docking   91

6.2.10   Advanced Sampling Techniques   91

6.2.11   Validation and Benchmarking of Flexible Docking   91

6.2.12   Integration with Experimental Data   91

6.2.13   Applications in Drug Discovery   92

6.2.14   Future Directions in Flexible Docking   92

7   Applications   93

7.1   Drug Discovery   93

7.1.1   Significance of Molecular Docking in Drug Discovery . . .   93

7.1.2   Mathematical Formulation of Binding Affinity   94

7.1.3   Examples of Successful Drug Discovery Through Docking   94

7.1.4   Challenges in Molecular Docking for Drug Discovery . . .   95

7.1.5   Integration with Experimental Validation   95

7.1.6   Combining Docking with ADMET Prediction   95

7.1.7   Application of Machine Learning in Drug Discovery . . .   95

7.1.8   Case Study: Designing a Kinase Inhibitor   96

7.1.9   Structure-Based Drug Design   96

7.1.10   Fragment-Based Drug Design   96

7.1.11   Future Directions in Drug Discovery with Docking   96

7.2   Virtual Screening   97

7.2.1   Principles of Virtual Screening   97

7.2.2   Mathematical Formulation of Screening Metrics   97

7.2.3   Methodologies in Virtual Screening   98

7.2.4   Applications of Virtual Screening   98

7.2.5   Case Study: Virtual Screening for Antiviral Agents   98

7.2.6   Integration with Experimental Validation   99

7.2.7   Challenges in Virtual Screening   99

7.2.8   Emerging Technologies in Virtual Screening   99

7.2.9   Future Directions in Virtual Screening   99

7.3   Case Studies 100

    7.3.1   Drug Repurposing: Rediscovering Famotidine 100

    7.3.2   Designing Selective Kinase Inhibitors 101

    7.3.3   Optimizing Antibody-Drug Conjugates 103

    7.3.4   Understanding Enzyme-Substrate Interactions: Trypsin-Catalyzed Reactions 105

    7.3.5   Targeting G-Protein Coupled Receptors (GPCRs) in Drug Discovery 107

    7.3.6   Virtual Screening for Anti-Malarial Compounds 109

    7.3.7   Rational Design of Antiviral Protease Inhibitors 111

    7.3.8   Optimizing Peptide Ligands for Receptor Binding 113

    7.3.9   Structure-Based Design of Anti-Inflammatory Agents . . 115

    7.3.10 Exploring Allosteric Modulation in Enzymes 117

    7.3.11 In Silico Design of Metalloenzyme Inhibitors 118

    7.3.12 Drug-Drug Interaction Studies: Cytochrome P450 Binding 120

    7.3.13 Prediction of Drug Metabolism: CYP2D6 Substrate Specificity 122

    7.3.14 Designing Antifungal Agents: Targeting Ergosterol Synthesis 125

    7.3.15 Optimizing Anti-HIV Protease Inhibitors 127

    7.3.16 Understanding Protein-Nucleic Acid Interactions: DNA Binding Proteins 129

    7.3.17 Advancements in Drug Delivery: Designing Nanocarriers . 132

    7.3.18 Integration of Machine Learning in Drug Design 134

8 Bioinformatics in Molecular Docking   137

8.1   Integration with Bioinformatics Tools 137

    8.1.1   Bioinformatics Databases and Molecular Docking 137

    8.1.2   Structural Bioinformatics for Target Identification 138

    8.1.3   Molecular Dynamics Simulations and Docking 138

    8.1.4   Bioinformatics Tools for Ligand Screening 138

8.1.5   Pathway Analysis and Systems Biology Integration 138

8.1.6   Genomic Data for Personalized Medicine 139

8.1.7   Integrating Network Pharmacology in Docking Studies . . 139

8.1.8   Metabolomics and Docking: Unraveling Metabolic Path-
          ways 139

8.1.9   Machine Learning for Predictive Docking Models 140

8.1.10   Bioinformatics Approaches for Solvent Effects 140

8.1.11   Data Integration for Systems Pharmacology 140

8.1.12   Examples of Bioinformatics-Integrated Docking Studies . 140

8.1.13   Challenges and Future Directions 141

8.2   Structural Bioinformatics 141

8.2.1   Protein Structure Determination 141

8.2.2   Homology Modeling for Target Prediction 143

8.2.3   Molecular Docking with Experimental Structures 145

8.2.4   Structural Bioinformatics in Virtual Screening 146

8.2.5   Protein-Ligand Interaction Analysis 148

8.2.6   Pharmacophore Modeling for Ligand Design 150

8.2.7   Integration of Protein Flexibility 151

8.2.8   Prediction of Binding Site Residues 153

8.2.9   Water Molecules in Docking Studies 154

8.2.10   Quantitative Structure-Activity Relationship 156

8.2.11   Understanding Allosteric Sites 157

8.2.12   Predicting Ligand-Induced Conformational Changes . . . 159

8.2.13   Examples of Structural Bioinformatics-Driven Docking
          Studies 161

8.2.14   Challenges and Future Directions 163

9   Challenges and Future Directions   165

9.1   Current Challenges 165

9.1.1   Protein Flexibility and Conformational Changes 165

9.1.2   Treatment of Solvent Effects 166

9.1.3 Handling Membrane Proteins 167

9.1.4 Scoring Function Accuracy 168

9.1.5 Treatment of Metal Ions and Cofactors 168

9.1.6 Incorporating Allosteric Interactions 169

9.1.7 Handling Large-Scale Virtual Screening 169

9.1.8 Accounting for Protein-Ligand Binding Pathways 170

9.1.9 Improving Predictions for Weak Binders 171

9.1.10 Handling Conformational Sampling 172

9.1.11 Incorporating Quantum Mechanical Effects 172

9.1.12 Addressing Ligand Flexibility 173

9.1.13 Data Standardization and Reproducibility 174

9.1.14 Enhancing User Accessibility 175

9.1.15 Integrating Experimental Data with Computational

Models 176

9.1.16 Addressing Bias in Training Datasets 177

9.1.17 Improving Predictive Capabilities for Challenging Targets 178

9.1.18 Incorporating Thermodynamic Considerations 179

9.1.19 Navigating Interactions in Dynamic Cellular Environments 180

9.1.20 Balancing Speed and Accuracy in Docking Calculations . 181

9.1.21 Promoting Open Science and Collaboration 182

9.2 Emerging Trends 183

9.2.1 Deep Learning in Scoring Functions 183

9.2.2 Hybrid Methods Integrating Experimental Data 184

9.2.3 Advancements in Quantum Mechanical Docking 185

9.2.4 Expanding Applications in Fragment-Based Drug Design 186

9.2.5 Incorporating Machine Learning in Ligand Design 187

9.2.6 GPU Acceleration for High-Performance Docking 188

9.2.7 Integration with Systems Biology 190

9.2.8 Enhanced Sampling Techniques for Conformational Ex-

ploration 190

9.2.9 3D Printing in Drug Design 191

9.2.10 Open Science Initiatives and Collaborative Platforms . . . 193

9.2.11 Interpretable AI Models for Decision Support 194

9.2.12 Quantifying Uncertainty in Predictions 195

9.2.13 Blockchain Technology for Data Security 196

9.2.14 Advances in Visualization Tools for Molecular Complexes 198

9.2.15 Multi-Target and Polypharmacology Considerations . . . 199

9.2.16 Ethical Considerations in AI-Driven Drug Discovery . . . 201

10 Appendix   **205**

10.1 Glossary 205

10.2 Abbreviations 207

10.3 References 209

Preface

Welcome to "Molecular Docking: A Formula Handbook." This handbook serves as a comprehensive guide for researchers, students, and professionals engaged in the exciting field of molecular docking. Whether you are a novice seeking an introduction to the basics or an experienced scientist looking for quick formula references, this handbook is designed to meet your needs.

Purpose of the Handbook

Molecular docking is a pivotal technique in computational chemistry and drug discovery. This handbook aims to provide a consolidated resource for various formulas, equations, and key concepts related to molecular docking. It is not only a quick reference for established researchers but also a learning companion for those entering the field.

Structure of the Handbook

The handbook is organized into several chapters, each focusing on a specific aspect of molecular docking. From the basics of ligand-receptor interactions to advanced topics like flexible docking and virtual screening, the handbook covers a wide range of essential concepts. The formula tables, examples, and illustrations aim to make complex ideas more accessible.

Who Should Read This Handbook

This handbook is suitable for:

- Researchers and scientists involved in molecular docking studies.

- Students studying computational chemistry, bioinformatics, or related fields.

- Professionals in the pharmaceutical industry engaged in drug discovery.

- Anyone interested in gaining insights into the molecular docking process.

How to Use This Handbook

Each chapter is designed to be self-contained, allowing readers to focus on specific topics of interest. You can use this handbook as a quick reference by jumping to relevant sections or as a comprehensive guide for in-depth learning.

Thank you for choosing "Molecular Docking: A Formula Handbook." May it serve as a valuable resource in your journey through the fascinating world of computational chemistry.

Chapter 1

Introduction

1.1 Background

Molecular docking is a crucial computational technique in the field of drug discovery, playing a pivotal role in understanding and predicting the interactions between small molecules (ligands) and target biomolecules (receptors). This methodology has gained prominence as a cost-effective and time-efficient means to explore the vast chemical space and identify potential drug candidates.

The roots of molecular docking can be traced back to the mid-20th century, with the advent of computational approaches to study molecular interactions. As the understanding of molecular biology deepened and computational resources expanded, molecular docking evolved into a sophisticated tool for simulating and predicting the binding modes of ligands to target proteins.

The fundamental concept behind molecular docking lies in predicting the three-dimensional structure of a ligand-receptor complex and assessing the strength of their interactions. Various forces, such as van der Waals interactions, hydrogen bonding, and electrostatic forces, contribute to the overall binding affinity. The scoring functions employed in molecular docking algorithms attempt to quantify these interactions mathematically.

$$\text{Binding Affinity} = a \cdot \text{Van der Waals Interactions} + b \qquad (1.1)$$
$$\cdot \text{Hydrogen Bonding} + c \cdot \text{Electrostatic Forces}$$

In this equation, the coefficients (a, b, c) are determined through empirical studies and optimization to enhance the accuracy of predicting binding affinities.

The success of molecular docking in drug discovery lies in its ability to rapidly evaluate a vast number of potential ligands against a target receptor. This accelerates the identification of lead compounds and aids in the design of novel therapeutic agents. The integration of molecular docking with experimental methods has become a standard approach in the pharmaceutical industry.

Advancements in computational power and algorithm development have further propelled molecular docking into the era of high-throughput virtual screening, enabling researchers to explore large chemical libraries efficiently. This has led to a paradigm shift in drug discovery, reducing the time and cost associated with traditional experimental screening.

The chapters that follow in this handbook will delve into the intricacies of molecular docking, covering computational methods, data preparation, validation techniques, and applications in drug discovery. The integration of bioinformatics tools, challenges, and future directions in the field will also be explored in detail.

1.2   Scope and Objectives

The scope of this handbook on molecular docking is comprehensive, aiming to provide an in-depth understanding of the theoretical foundations, computational methods, and practical applications in drug discovery. The objectives of this book include elucidating the principles underlying molecular docking techniques, exploring the diverse computational methods employed, and highlighting the challenges and future directions in this dynamic field.

1.2.1 Understanding Molecular Interactions

The primary goal is to help readers comprehend the intricacies of molecular interactions at the atomic and molecular levels. By exploring the principles of ligand-receptor binding, van der Waals forces, hydrogen bonding, and electrostatic interactions, readers will gain insights into the factors influencing molecular docking outcomes.

1.2.2 Computational Methods and Algorithms

This handbook delves into the various computational methods and algorithms employed in molecular docking. From grid-based approaches to advanced Monte Carlo and genetic algorithms, readers will gain a comprehensive understanding of the tools available for simulating and predicting ligand-receptor interactions.

1.2.3 Data Preparation and Validation Techniques

An essential aspect of molecular docking is the proper preparation of molecular data and the validation of docking results. This book addresses the critical steps involved in data preparation for ligands and receptors and introduces validation metrics and benchmark datasets to assess the accuracy of docking predictions.

1.2.4 Applications in Drug Discovery

Molecular docking plays a crucial role in drug discovery, providing insights into the interactions between small molecules (ligands) and target proteins. This subsection explores the diverse applications of molecular docking in the drug development process.

Lead Identification and Optimization

One of the primary applications is the identification and optimization of lead compounds. Molecular docking helps screen large chemical libraries to identify potential lead molecules that exhibit favorable binding interactions with the target protein. Subsequent optimization involves refining these leads to enhance their binding affinity and specificity.

Virtual Screening

Virtual screening is a powerful application where molecular docking is used to assess the binding affinity of a vast number of compounds against a target protein. This high-throughput screening aids in prioritizing potential drug candidates, saving time and resources in the initial stages of drug discovery.

Polypharmacology Studies

Understanding the polypharmacology of drug candidates is crucial for their success. Molecular docking allows researchers to explore how a candidate molecule interacts with multiple target proteins, providing insights into potential off-target effects and enabling the design of multi-target drugs.

ADMET Prediction

In silico ADMET (Absorption, Distribution, Metabolism, Excretion, and Toxicity) prediction is another application facilitated by molecular docking. By analyzing the binding characteristics, researchers can infer crucial pharmacokinetic properties, helping eliminate compounds with unfavorable ADMET profiles early in the drug discovery process.

Protein-Protein Interaction Inhibition

Molecular docking extends beyond small molecule binding to single proteins. It is increasingly employed in studying and designing compounds that disrupt protein-protein interactions, a challenging but promising avenue in drug discovery, particularly for diseases with complex molecular pathways.

These applications highlight the versatility and impact of molecular docking in streamlining the drug discovery pipeline. The integration of computational tools like molecular docking has revolutionized the field, contributing to the development of safer and more effective therapeutic agents.

1.2.5 Integration with Bioinformatics Tools

The integration of molecular docking with bioinformatics tools enhances the overall analysis and interpretation of complex biological data. This subsection explores the synergies between molecular docking techniques and bioinformatics, providing a holistic approach to computational biology.

Structural Bioinformatics

Structural bioinformatics plays a crucial role in understanding the three-dimensional structures of biological macromolecules. Integrating molecular docking with structural bioinformatics tools allows researchers to analyze and visualize the binding interactions in the context of protein structures. This aids in identifying key residues, binding pockets, and structural features influencing ligand binding.

Network Analysis

Network analysis tools in bioinformatics provide a systemic view of molecular interactions within a biological system. By integrating molecular docking results into network analyses, researchers can uncover intricate relationships between proteins, identify key nodes in signaling pathways, and prioritize targets for drug development.

Pathway Analysis

Understanding the impact of ligand binding on cellular pathways is essential. Bioinformatics tools for pathway analysis can be integrated with molecular docking data to assess the broader implications of ligand-protein interactions. This integration contributes to a more comprehensive understanding of the biological consequences of drug binding.

Data Mining and Machine Learning

Bioinformatics leverages data mining and machine learning techniques for pattern recognition and prediction. Integration with molecular docking results enables the development of predictive models for ligand binding affinity, helping prioritize potential drug candidates for experimental validation.

Visualization Tools

Visualization tools in bioinformatics enhance the interpretation of complex molecular interactions. Integrating molecular docking results with advanced visualization tools allows for the creation of interactive, 3D representations of ligand-protein complexes. This aids in conveying structural insights to a broader audience, facilitating collaboration and knowledge dissemination.

The integration of molecular docking with bioinformatics tools opens avenues for a more holistic and systems-level understanding of biological processes. Researchers can leverage these integrated approaches to unravel intricate molecular mechanisms, identify novel drug targets, and accelerate drug discovery efforts.

1.2.6   Challenges and Future Directions

The field of molecular docking, while making significant strides, faces several challenges and holds exciting possibilities for future advancements. This subsection delves into the current challenges and outlines potential directions for future research.

Scoring Function Accuracy

One of the persistent challenges in molecular docking is improving the accuracy of scoring functions. Despite advancements, accurately predicting the binding affinity between ligands and receptors remains a complex task. Future research may focus on developing more reliable scoring functions, incorporating machine learning approaches to enhance predictive accuracy.

Incorporating Flexibility

Capturing the flexibility of biological macromolecules, such as proteins and nucleic acids, poses a challenge in molecular docking. Future directions may involve the development of advanced algorithms and techniques that can effectively handle the inherent flexibility of biomolecular structures, enabling more realistic ligand binding predictions.

Treatment of Water Molecules

The accurate treatment of water molecules in molecular docking simulations is essential for realistic binding predictions. Current methods often simplify the role of water, but future research could explore more sophisticated approaches to account for water-mediated interactions and the dynamic behavior of solvent molecules in the binding site.

Handling Macromolecular Complexes

Molecular docking traditionally focuses on binary interactions between a ligand and a receptor. Future directions may involve extending these techniques to handle complex scenarios involving multiple proteins, nucleic acids, and ligands. This expansion is crucial for understanding intricate cellular processes and developing therapeutics targeting complex molecular assemblies.

Integrating Experimental Data

While computational methods like molecular docking provide valuable insights, integrating experimental data remains a challenge. Future research may aim at developing frameworks that seamlessly combine computational predictions with experimental results, ensuring a more comprehensive and accurate understanding of molecular interactions.

Exploring Novel Applications

The future of molecular docking extends beyond traditional drug discovery applications. Researchers may explore novel applications, such as predicting protein-protein interactions, designing bio-inspired materials, and contributing to the fields of synthetic biology and personalized medicine.

Chapter 2

Basics of Molecular Docking

2.1 Molecular Docking Overview

Molecular docking is a computational method used in drug discovery and structural biology to predict the interactions between a small molecule (ligand) and a target biomolecule (receptor). The primary objective is to simulate the binding process and predict the most favorable orientation and conformation of the ligand within the binding site of the receptor. This overview explores the key concepts and fundamental principles that underpin molecular docking.

2.1.1 Binding Affinity and Energetics

Central to molecular docking is the concept of binding affinity, representing the strength of the interaction between the ligand and receptor. The binding affinity is determined by the sum of various energy contributions, including van der Waals forces, electrostatic interactions, hydrogen bonding, and desolvation effects. The overall energetics of the binding process can be mathematically expressed as:

$$
\begin{aligned}
\text{Binding Energy} = {} & \text{Van der Waals Energy} + \text{Electrostatic Energy} \\
& + \text{Hydrogen Bonding Energy} - \text{Desolvation Energy}
\end{aligned}
\tag{2.1}
$$

Understanding these energy components is crucial for predicting and optimizing ligand-receptor interactions.

2.1.2   Docking Algorithm Types

Molecular docking employs various algorithms to predict the binding interactions between ligands and receptors accurately. This subsection provides an overview of different types of docking algorithms, each designed to address specific aspects of ligand-receptor interactions.

Geometric Matching Algorithms

Geometric matching algorithms are among the earliest docking methods. They operate based on the geometric complementarity of ligands and receptors. One well-known algorithm is the Fast Fourier Transform (FFT)-based method, where ligand and receptor shapes are matched in three-dimensional space to identify potential binding sites.

Grid-Based Algorithms

Grid-based algorithms divide the docking space into a grid, simplifying the search for favorable binding positions. The Lamarckian Genetic Algorithm and Solis-Wets local search algorithm are examples. These algorithms explore various ligand conformations within the grid, optimizing ligand poses for optimal binding.

Monte Carlo Methods

Monte Carlo methods simulate the random movements of a ligand within the binding site to explore potential binding configurations. These algorithms, such as Metropolis Monte Carlo, iteratively sample ligand conformations and evaluate their binding energies, gradually converging to energetically favorable poses.

Molecular Dynamics-Based Algorithms

Molecular Dynamics (MD) simulations simulate the dynamic behavior of ligands and receptors over time. While computationally intensive, MD-based docking algorithms provide valuable insights into the flexibility and dynamics of the binding process. Integrating MD with docking enhances the accuracy of predictions.

Machine Learning-Based Algorithms

Recent advancements in machine learning have led to the development of docking algorithms that leverage predictive models. These algorithms, often trained on large datasets of known ligand-receptor complexes, can rapidly predict binding affinities and explore ligand conformations efficiently.

Hybrid Approaches

Hybrid docking approaches combine multiple algorithms to capitalize on their individual strengths. For example, a docking protocol might incorporate geometric matching for an initial search, followed by refinement using a molecular dynamics-based algorithm. Hybrid approaches aim to improve accuracy and efficiency.

Mathematical Formulation:

The scoring function used in docking algorithms often involves mathematical expressions to evaluate the binding energy of a given ligand-receptor complex. One common form is:

$$E_{\text{bind}} = E_{\text{vdW}} + E_{\text{elec}} + E_{\text{desolv}} + E_{\text{other}} \tag{2.2}$$

Where:

E_{vdW} is the van der Waals energy,

E_{elec} is the electrostatic energy,

E_{desolv} is the desolvation energy,

E_{other} includes additional terms.

Chemical Formula:

In molecular docking, ligands and receptors are represented using chemical formulas. For instance, a ligand with a chemical formula of a small organic molecule might be denoted as $C_nH_mO_p$, where n, m, and p represent the respective numbers of carbon, hydrogen, and oxygen atoms.

Understanding the strengths and limitations of different docking algorithm types is crucial for selecting the most suitable method based on the specific characteristics of the ligands and receptors under investigation.

2.1.3 Scoring Functions

Scoring functions play a pivotal role in molecular docking, quantifying the affinity between ligands and receptors. This subsection explores different scoring functions used to evaluate binding energies and predict the most favorable ligand-receptor complexes.

Empirical Scoring Functions

Empirical scoring functions are rule-based and rely on predefined parameters to estimate binding affinities. One common example is the Generalized Born and Surface Area (GB/SA) method. The empirical scoring function E_{emp} can be expressed as:

$$E_{\text{emp}} = aE_{\text{vdW}} + bE_{\text{elec}} + cE_{\text{desolv}} + dE_{\text{other}} \qquad (2.3)$$

Here, E_{vdW}, E_{elec}, E_{desolv}, and E_{other} represent van der Waals, electrostatic, desolvation, and additional energy terms, respectively. The coefficients a, b, c, and d are parameters fitted to experimental data.

Force Field-Based Scoring Functions

Force field-based scoring functions utilize molecular mechanics force fields to calculate the potential energy of a ligand-receptor complex. The scoring function E_{ff} is given by:

$$E_{ff} = E_{bonded} + E_{nonbonded} + E_{electrostatic} + E_{van\ der\ Waals} \qquad (2.4)$$

The terms include bonded interactions (bonds, angles, dihedrals), nonbonded interactions (van der Waals and electrostatic), and additional terms.

Knowledge-Based Scoring Functions

Knowledge-based scoring functions derive information from experimental databases to predict binding affinities. The potential energy E_{kb} is calculated based on statistical analyses of known ligand-receptor complexes:

$$E_{kb} = -k \log P(\text{complex}) \qquad (2.5)$$

Here, $P(\text{complex})$ represents the probability of observing the given ligand-receptor complex in the database, and k is a scaling factor.

Chemical Formula:

Chemical formulas are essential for representing ligands involved in scoring functions. For example, a ligand with the chemical formula $C_n H_m O_p$ denotes the number of carbon (n), hydrogen (m), and oxygen (p) atoms, providing insights into the molecular structure contributing to the scoring function.

Understanding the principles behind these scoring functions enhances the interpretation of molecular docking results and aids in selecting appropriate methods for specific research objectives.

2.1.4   Ligand and Receptor Preparation

The accurate preparation of ligands and receptors is crucial for reliable molecular docking studies. This subsection discusses key steps involved in the preparation process, including mathematical formulations and chemical representations.

Ligand Preparation

Ligand preparation involves optimizing the ligand structure for docking simulations. A common step is energy minimization, aiming to achieve a stable and realistic conformation. The energy minimization process can be represented mathematically as follows:

$$E_{\min} = \underset{\text{conformation}}{\text{minimize}} \; (E_{\text{ligand}} + E_{\text{constraints}}) \tag{2.6}$$

Here, E_{ligand} is the potential energy of the ligand, and $E_{\text{constraints}}$ includes any applied constraints during the optimization.

Chemical formulas are essential for representing ligands, and a simplified example can be given by the chemical formula of a small organic molecule: $C_n H_m O_p$, where n, m, and p represent the respective numbers of carbon, hydrogen, and oxygen atoms.

Receptor Preparation

Receptor preparation involves optimizing the receptor structure to create a realistic binding site for ligand interactions. This often includes removing water molecules, adding hydrogen atoms, and optimizing side-chain conformations. Mathematically, receptor preparation can be expressed as:

$$E_{\min} = \underset{\text{conformation}}{\text{minimize}} \; (E_{\text{receptor}} + E_{\text{constraints}}) \tag{2.7}$$

Here, E_{receptor} is the potential energy of the receptor, and $E_{\text{constraints}}$ includes any applied constraints during the optimization.

Chemical formulas are also crucial for representing receptors. For example, the chemical formula of a protein could be denoted as $C_n H_m O_p N_q S_r$, including carbon, hydrogen, oxygen, nitrogen, and sulfur atoms.

Understanding the intricacies of ligand and receptor preparation is fundamental for ensuring the reliability of molecular docking simulations and obtaining meaningful results.

2.1.5 Visualization of Docking Results

Visualizing docking results is crucial for interpreting and communicating findings effectively. This subsection explores various methods for visualizing molecular docking results, incorporating both mathematical formulations and chemical representations.

3D Visualization

Three-dimensional (3D) visualization is a powerful method for representing ligand-receptor complexes. The spatial arrangement of atoms can be mathematically described using Cartesian coordinates. For a ligand with N atoms, the 3D coordinates can be denoted as (x_i, y_i, z_i), where $i = 1, 2, \ldots, N$.

Binding Affinity Heatmaps

Binding affinity heatmaps provide a visual representation of ligand-receptor interactions. The heatmap matrix M can be formulated based on the binding affinities E_{bind} between ligands and receptors:

$$M_{ij} = E_{\text{bind}}(L_i, R_j) \tag{2.8}$$

Here, L_i represents the i-th ligand, R_j represents the j-th receptor, and E_{bind} is the binding affinity.

2D Interaction Diagrams

Two-dimensional (2D) interaction diagrams illustrate specific interactions between ligands and receptors. Mathematically, these interactions can be represented as:

$$\text{Interaction}_{ij} = \text{Type}_{ij} + \text{Distance}_{ij} + \text{Angle}_{ij} \qquad (2.9)$$

Here, Type_{ij} denotes the type of interaction (e.g., hydrogen bonding), Distance_{ij} represents the distance between interacting atoms, and Angle_{ij} indicates the angle of interaction.

Chemical Formulas for Ligand-Receptor Complexes

Chemical formulas remain essential for conveying the molecular composition of ligand-receptor complexes. For instance, the chemical formula $C_n H_m O_p N_q$ represents a complex containing carbon, hydrogen, oxygen, and nitrogen atoms.

Understanding these visualization techniques enhances the interpretation of docking results, facilitating the identification of key interactions and the overall binding landscape.

2.1.6   Challenges and Limitations

Despite its utility, molecular docking is not without challenges. The method may encounter issues in accurately predicting ligand binding modes, especially in cases of significant conformational changes or flexible binding sites. Additionally, scoring functions may have limitations in accurately capturing the complex thermodynamics of ligand binding.

2.1.7   Applications in Drug Discovery

Molecular docking plays a crucial role in drug discovery, aiding researchers in identifying potential drug candidates and understanding their interactions with target receptors. This subsection explores various applications of molecular docking in drug discovery, incorporating both mathematical formulations and chemical representations.

Virtual Screening

Virtual screening is a prominent application where molecular docking is employed to screen large compound libraries and identify potential drug candidates. The process involves evaluating the binding affinity (E_{bind}) of ligands to target receptors. Mathematically, virtual screening can be expressed as:

$$\text{Virtual Screening Score}_i = E_{\text{bind}}(L_i, R_{\text{target}}) \qquad (2.10)$$

Here, L_i represents the i-th ligand, and R_{target} is the target receptor.

Lead Optimization

Molecular docking is instrumental in lead optimization, refining initial hits into more potent drug candidates. Optimization involves modifying ligand structures to enhance binding affinity. Mathematically, lead optimization can be represented as:

$$\text{Optimized Binding Affinity}_i = \underset{\text{modifications}}{\text{maximize}} \ (E_{\text{bind}}(L_i, R_{\text{target}})) \qquad (2.11)$$

ADME/T Properties Prediction

Absorption, Distribution, Metabolism, Excretion, and Toxicity (ADME/T) properties are crucial considerations in drug development. Molecular docking aids in predicting these properties based on ligand-receptor interactions. The prediction can be formulated as:

$$\text{ADME/T Prediction}_i = f(E_{\text{bind}}(L_i, R_{\text{target}})) \qquad (2.12)$$

Here, f is a function mapping binding affinity to ADME/T properties.

Chemical Formulas for Drug Candidates

Chemical formulas are essential for representing potential drug candidates. For example, a drug candidate with the chemical formula $C_n H_m O_p N_q$ denotes the molecular composition of carbon, hydrogen, oxygen, and nitrogen atoms.

Understanding these applications in drug discovery empowers researchers to leverage molecular docking as a valuable tool in the development of novel therapeutics.

2.1.8   Case Studies

This subsection presents case studies showcasing real-world applications of molecular docking, accompanied by relevant mathematical formulations and chemical representations.

Case Study 1: Drug Repurposing

In drug repurposing, existing drugs are investigated for new therapeutic uses. Molecular docking facilitates the exploration of potential interactions between approved drugs and different target receptors. The binding affinity (E_{bind}) in this context is crucial:

$$\text{Binding Affinity}_{\text{repurpose}} = E_{\text{bind}}(L_{\text{existing drug}}, R_{\text{new target}}) \qquad (2.13)$$

Here, $L_{\text{existing drug}}$ is an approved drug, and $R_{\text{new target}}$ is the newly identified target receptor.

Case Study 2: Protein-Protein Interactions

Molecular docking is not limited to small molecule interactions; it is also employed in studying protein-protein interactions. The binding energy (E_{bind}) between two proteins can be expressed mathematically:

$$\text{Binding Energy}_{\text{protein-protein}} = E_{\text{bind}}(P_1, P_2) \qquad (2.14)$$

Here, P_1 and P_2 represent the two interacting proteins.

Case Study 3: Enzyme Substrate Interactions

Understanding enzyme-substrate interactions is crucial for drug design. Molecular docking aids in studying the binding affinity (E_{bind}) between enzymes and substrates:

$$\text{Enzyme-Substrate Binding Affinity} = E_{bind}(E_{enzyme}, S_{substrate}) \qquad (2.15)$$

Here, E_{enzyme} represents the enzyme, and $S_{substrate}$ represents the substrate.

Chemical Formulas for Investigated Molecules

Chemical formulas play a key role in representing the molecular composition of investigated molecules. For instance, the chemical formula $C_n H_m O_p N_q$ illustrates the combination of carbon, hydrogen, oxygen, and nitrogen atoms.

By delving into these case studies, researchers can gain insights into the diverse applications of molecular docking across different areas of molecular biology and drug discovery.

2.1.9 Emerging Trends

As the field evolves, emerging trends in molecular docking include the integration of machine learning techniques for improved accuracy and efficiency. Artificial intelligence-driven approaches contribute to enhancing the predictive power of docking simulations, guiding researchers toward more successful drug discovery outcomes.

2.2 Key Concepts

This section delves into fundamental concepts in molecular docking, providing a comprehensive understanding of the key principles that underpin this computational technique. The discussion includes both mathematical formulations and chemical representations.

2.2.1   Binding Affinity

Binding affinity (E_{bind}) is a central concept in molecular docking, representing the strength of interaction between a ligand and a receptor. It is mathematically expressed as:

$$E_{\text{bind}}(L, R) = E_{\text{ligand}} + E_{\text{receptor}} + E_{\text{interactions}} \qquad (2.16)$$

Here, E_{ligand}, E_{receptor}, and $E_{\text{interactions}}$ denote the energy contributions from the ligand, receptor, and their interactions, respectively.

2.2.2   Scoring Functions

Scoring functions play a crucial role in evaluating the binding affinity between a ligand and a receptor. One common scoring function is the empirical scoring function, which combines various energy terms:

$$\text{Empirical Score} = a \cdot E_{\text{vdW}} + b \cdot E_{\text{elec}} + c \cdot E_{\text{desolv}} + \ldots \qquad (2.17)$$

Here, E_{vdW}, E_{elec}, and E_{desolv} represent van der Waals, electrostatic, and desolvation energies, respectively.

2.2.3   Docking Algorithm

The docking algorithm is a critical component in molecular docking, determining the spatial arrangement of ligands within the binding site of a receptor. Several algorithms exist, each with its unique approach to exploring the conformational space of ligands. Here, we discuss the Lamarckian Genetic Algorithm (LGA) as an example, outlining its steps and key mathematical aspects.

Lamarckian Genetic Algorithm (LGA)

The LGA is an iterative optimization algorithm that simulates the natural process of evolution. It is employed to search for energetically favorable ligand conformations within the binding site of a receptor.

$$\text{LGA Steps: } \quad \text{Selection} \rightarrow \text{Crossover} \rightarrow \text{Mutation} \rightarrow \text{Evaluation} \rightarrow \text{Acceptance} \tag{2.18}$$

1. **Selection:** In this step, individuals (ligand conformations) are selected based on their fitness, which is determined by their binding affinity. High-affinity conformations are more likely to be selected for the next generation.

2. **Crossover:** Crossover involves combining genetic material from two parent conformations to generate new offspring conformations. This mimics the genetic recombination observed in natural evolution.

3. **Mutation:** Random changes are introduced into the genetic material of selected individuals, leading to the exploration of new conformations. This helps the algorithm escape local energy minima.

4. **Evaluation:** The fitness of the newly generated conformations is evaluated based on the scoring function. The goal is to identify conformations with high binding affinity.

5. **Acceptance:** The newly generated conformations are accepted or rejected based on their fitness. This step influences the composition of the next generation.

Mathematical Formula

The scoring function used in the evaluation step often combines various energy terms, contributing to the overall fitness of a conformation:

$$\text{Fitness} = a \cdot E_{\text{vdW}} + b \cdot E_{\text{elec}} + c \cdot E_{\text{desolv}} + \ldots \tag{2.19}$$

Here, E_{vdW}, E_{elec}, and E_{desolv} represent van der Waals, electrostatic, and desolvation energies, respectively.

Understanding the docking algorithm and its mathematical aspects is crucial for researchers employing molecular docking in their studies.

2.2.4   Ligand and Receptor Representations

The representation of ligands and receptors is a crucial aspect of molecular docking, influencing how these molecules are modeled and interact within the computational framework. Here, we explore different ways to represent ligands and receptors, including mathematical and chemical formulations.

Mathematical Formulas

Mathematical formulas can be used to represent the structure and properties of ligands and receptors. For instance, a ligand may be described using a molecular formula, denoting the number of atoms of each element. Consider a ligand with the chemical formula $C_n H_m O_p N_q$, indicating the presence of carbon (C), hydrogen (H), oxygen (O), and nitrogen (N) atoms in specific quantities.

Chemical Formulas

Chemical formulas provide a concise representation of the atomic composition of molecules. Ligands and receptors can be depicted using structural formulas, such as:

$$\text{Ligand:}\quad CH_3 - CH_2 - CH_2 - CH_2 - OH \tag{2.20}$$

This represents a simple alcohol molecule. Similarly, a receptor's chemical structure can be depicted using chemical formulas, providing insights into its molecular architecture.

Molecular Graphics

Visualization tools play a crucial role in understanding ligand-receptor interactions. Molecular graphics, such as three-dimensional molecular models, can visually represent the spatial arrangement of atoms and bonds. Visualization aids in comprehending the conformation and orientation of ligands within the binding site.

Intermolecular Interactions

Understanding the intermolecular interactions between ligands and receptors is essential. These interactions include van der Waals forces, hydrogen bonding, and electrostatic interactions. Mathematically, these interactions can be expressed through force field equations that describe the energy contributions from different interaction types.

$$E_{\text{interactions}} = E_{\text{vdW}} + E_{\text{HB}} + E_{\text{elec}} + \dots \tag{2.21}$$

Here, E_{vdW} represents van der Waals energy, E_{HB} represents hydrogen bonding energy, and E_{elec} represents electrostatic energy.

By exploring these representations, researchers can gain a comprehensive understanding of the structural and energetic aspects of ligand-receptor interactions in the context of molecular docking.

Chapter 3

Computational Methods

3.1 Force Fields

Force fields are indispensable tools in molecular docking, providing a quantitative framework for describing the interactions between atoms in a molecular system. These computational models simulate the potential energy of a molecular system, incorporating various parameters to represent bond stretching, angle bending, and torsional rotations. In this section, we explore the fundamentals of force fields and their crucial role in molecular docking simulations.

3.1.1 The Foundation of Force Fields

Force fields serve as the mathematical framework for describing the interactions between atoms and molecules in molecular docking simulations. These fields are crucial for calculating the potential energy of a molecular system, guiding the exploration of conformational space. Here, we delve into the foundational aspects of force fields, including mathematical formulations and chemical considerations.

Mathematical Formulas

The potential energy (E) of a molecular system in a force field is often expressed as the sum of various energy terms:

$$E = E_{\text{bond}} + E_{\text{angle}} + E_{\text{dihedral}} + E_{\text{vdW}} + E_{\text{elec}} + \dots \qquad (3.1)$$

Each term represents a specific type of interaction: bond stretching (E_{bond}), angle bending (E_{angle}), dihedral angle rotation (E_{dihedral}), van der Waals interactions (E_{vdW}), electrostatic interactions (E_{elec}), and more.

Chemical Formulas

The parameters in force fields are derived from experimental data and quantum mechanical calculations. For example, bond parameters (K_b, b_0) for a simple harmonic potential in the bond term can be assigned based on experimental bond lengths:

$$E_{\text{bond}} = \sum_{\text{bonds}} K_b (r - b_0)^2 \qquad (3.2)$$

Here, r is the current bond length, b_0 is the equilibrium bond length, and K_b is the force constant.

Force Field Implementation

Force fields are implemented in molecular docking software to calculate the energy of a given conformation. The implementation involves assigning force field parameters to each atom type, defining bond connectivity, and specifying interaction potentials.

Parameterization Challenges

Accurate force field parameterization remains a challenge, as it requires balancing the accuracy of simulations with computational efficiency. Researchers continually refine force field parameters to better reproduce experimental observations.

Understanding the mathematical foundation of force fields is essential for researchers engaged in molecular docking studies, as it directly influences the accuracy and reliability of simulations.

3.1.2 Bonded Interactions: Springs and Angles

Bonded interactions, such as springs representing bonds and angles, play a crucial role in force fields, contributing to the overall potential energy of a molecular system. In this section, we explore the mathematical formulations and chemical considerations associated with bonded interactions.

Bond Stretching (Springs)

The stretching of chemical bonds can be modeled as springs. The potential energy (E_{bond}) associated with bond stretching is often described using Hooke's law:

$$E_{\text{bond}} = \sum_{\text{bonds}} \frac{1}{2} K_b (r - b_0)^2 \tag{3.3}$$

Here, r is the current bond length, b_0 is the equilibrium bond length, K_b is the force constant, and the summation is over all bonded pairs of atoms.

Angle Bending

Angle bending, representing the deformation of chemical angles, is another bonded interaction term. The potential energy (E_{angle}) associated with angle bending is often modeled as:

$$E_{\text{angle}} = \sum_{\text{angles}} \frac{1}{2} K_\theta (\theta - \theta_0)^2 \tag{3.4}$$

Here, θ is the current angle, θ_0 is the equilibrium angle, K_θ is the force constant, and the summation is over all angles in the molecular system.

Chemical Formulas

In chemical formulas, bonded interactions can be represented by specific bond types and angles. For example, a chemical bond between atoms A and B can be denoted as $A - B$, and the corresponding equilibrium bond length as $b_{0,\mathrm{AB}}$. Similarly, an angle formed by atoms A, B, and C can be represented as $\angle ABC$, with the equilibrium angle denoted as $\theta_{0,\mathrm{ABC}}$.

Force Field Parameters

Force field parameters, such as bond force constants and equilibrium values, are determined through fitting to experimental data or quantum mechanical calculations. These parameters play a crucial role in accurately describing the behavior of bonded interactions within a molecular system.

Understanding the mathematical representations of bonded interactions is essential for researchers involved in molecular docking simulations, influencing the accuracy and reliability of the computational models.

3.1.3 Non-Bonded Interactions: Van der Waals and Electrostatic Forces

Non-bonded interactions, specifically Van der Waals (vdW) forces and electrostatic forces, are integral components of force fields governing molecular docking simulations. In this section, we explore the mathematical formulations and chemical considerations associated with these non-bonded interactions.

Van der Waals Interactions

Van der Waals forces arise from the attractive and repulsive interactions between atoms due to induced dipoles. The potential energy (E_{vdW}) associated with Van der Waals interactions is often modeled using Lennard-Jones potential:

$$E_{\mathrm{vdW}} = \sum_{\mathrm{pairs}} \left[4\epsilon \left(\left(\frac{\sigma}{r}\right)^{12} - \left(\frac{\sigma}{r}\right)^{6} \right) \right] \tag{3.5}$$

Here, ϵ is the depth of the potential well, σ is the finite distance at which the inter-particle potential is zero, r is the distance between atom pairs, and the summation is over all non-bonded atom pairs.

Electrostatic Interactions

Electrostatic forces arise from the interaction of charged particles. The potential energy (E_{elec}) associated with electrostatic interactions is modeled using Coulomb's law:

$$E_{\text{elec}} = \sum_{\text{pairs}} \frac{q_i q_j}{4\pi\epsilon_0 r_{ij}} \tag{3.6}$$

Here, q_i and q_j are the charges of atoms i and j, ϵ_0 is the vacuum permittivity, r_{ij} is the distance between the charges, and the summation is over all non-bonded atom pairs.

Chemical Formulas

In chemical formulas, non-bonded interactions are represented by atoms and their distances. For example, a non-bonded interaction between atoms A and B can be denoted as $A \ldots B$, representing the spatial proximity without chemical bonding.

Force Field Parameters

Force field parameters for Van der Waals and electrostatic interactions include atom-specific parameters such as ϵ, σ, and partial charges (q). These parameters are crucial for accurately capturing the non-bonded interactions within a molecular system.

Understanding the mathematical representations of non-bonded interactions is essential for researchers involved in molecular docking simulations, influencing the accuracy and reliability of the computational models.

3.1.4   Combining Forces: The Total Potential Energy

In molecular docking simulations, the total potential energy of a system is the sum of various energy components, including bonded interactions, Van der Waals forces, and electrostatic interactions. The mathematical expression for the total potential energy (E_{total}) is given by:

$$E_{\text{total}} = E_{\text{bond}} + E_{\text{angle}} + E_{\text{vdW}} + E_{\text{elec}} \tag{3.7}$$

Here, E_{bond}, E_{angle}, E_{vdW}, and E_{elec} represent the contributions from bonded interactions, angle bending, Van der Waals forces, and electrostatic interactions, respectively.

This equation captures the comprehensive interplay of different forces governing the behavior of molecules during docking simulations. The force field parameters for each term in the equation are determined based on experimental data or quantum mechanical calculations.

Chemical Formulas

In chemical formulas, the total potential energy is a sum of the individual energy components. For example, in a molecular system with atoms A, B, and C, the total potential energy can be represented as:

$$A - B + \angle ABC + A \dots B + \frac{q_A q_B}{4\pi\epsilon_0 r_{AB}} \tag{3.8}$$

This representation highlights the combination of bonded and non-bonded interactions contributing to the overall potential energy.

Understanding the total potential energy is crucial for researchers in assessing the stability and energetics of molecular complexes during docking simulations. It forms the basis for evaluating the feasibility and reliability of docking results.

3.1.5   Popular Force Fields in Molecular Docking

Several force fields are widely used in molecular docking simulations. AMBER (Assisted Model Building with Energy Refinement), CHARMM (Chemistry at HARvard Molecular Mechanics), and GROMOS (GROningen MOlecular Simulation) are among the most recognized force fields. Each force field has its unique set of parameters and approximations, making them suitable for different types of molecular systems.

3.1.6   Parameterization of Force Fields

The accurate representation of molecular interactions in docking simulations relies on the parameterization of force fields. Force field parameters, such as bond lengths, bond angles, and non-bonded interactions, are crucial for capturing the realistic behavior of molecules. In this section, we delve into the mathematical formulations and chemical considerations associated with force field parameterization.

Bond Parameters

The bond parameters in a force field include equilibrium bond length (r_0) and force constant (k). The potential energy associated with bonded interactions (E_{bond}) can be expressed as:

$$E_{\text{bond}} = \sum_{\text{bonds}} \frac{1}{2} k_i (r_i - r_{0i})^2 \tag{3.9}$$

Here, r_i is the current bond length, r_{0i} is the equilibrium bond length, k_i is the force constant, and the summation is over all bonded interactions.

Angle Parameters

Angle parameters in a force field include equilibrium angle (θ_0) and force constant (k_θ). The potential energy associated with angle bending (E_{angle}) is given by:

$$E_{\text{angle}} = \sum_{\text{angles}} \frac{1}{2} k_{\theta_i} (\theta_i - \theta_{0i})^2 \tag{3.10}$$

Here, θ_i is the current angle, θ_{0i} is the equilibrium angle, k_{θ_i} is the force constant, and the summation is over all angle bending interactions.

Non-Bonded Parameters

Non-bonded interactions involve Van der Waals and electrostatic forces. The parameters include Van der Waals parameters (ϵ, σ) and partial charges (q). The potential energy associated with Van der Waals (E_{vdW}) and electrostatic (E_{elec}) interactions were discussed in previous sections.

Chemical Formulas

In chemical formulas, force field parameters are represented as specific values associated with atoms and their interactions. For example, a bond between atoms A and B with equilibrium length r_{0AB} and force constant k_{AB} can be denoted as $A - B(r_{0AB}, k_{AB})$.

Parameterization of force fields is a critical step in molecular docking simulations, ensuring the accuracy of predicted molecular interactions.

3.1.7   Validation and Benchmarking

Validation and benchmarking are essential steps in assessing the accuracy and reliability of molecular docking methods. These processes involve comparing predicted results with experimental data or established benchmarks to ensure the robustness of the methodology.

Validation Metrics

To quantify the performance of a docking algorithm, various validation metrics are employed. One commonly used metric is the Root Mean Square Deviation (RMSD), which measures the deviation between predicted and experimental structures:

$$\text{RMSD} = \sqrt{\frac{1}{N}\sum_{i=1}^{N}(x_i - x_i')^2} \tag{3.11}$$

Here, N is the number of atoms, x_i are the coordinates of atom i in the experimental structure, and x_i' are the coordinates in the predicted structure.

Benchmark Datasets

Benchmark datasets are curated sets of molecular complexes with known experimental structures and binding affinities. These datasets serve as a standardized evaluation tool for docking algorithms. The potential energy (E_{pred}) predicted by a docking algorithm can be compared to the experimental binding affinity (E_{exp}) using metrics such as the correlation coefficient (R^2):

$$R^2 = \frac{\left(\sum_{i=1}^{N}(E_{\text{pred},i} - \bar{E}_{\text{pred}})(E_{\text{exp},i} - \bar{E}_{\text{exp}})\right)^2}{\left(\sum_{i=1}^{N}(E_{\text{pred},i} - \bar{E}_{\text{pred}})^2 \sum_{i=1}^{N}(E_{\text{exp},i} - \bar{E}_{\text{exp}})^2\right)} \tag{3.12}$$

Here, N is the number of complexes, $E_{\text{pred},i}$ and $E_{\text{exp},i}$ are the predicted and experimental energies for complex i, and $\bar{E}_{\text{pred}}$ and $\bar{E}_{\text{exp}}$ are the respective means.

Chemical Formulas

In chemical formulas, the comparison between predicted and experimental results can be represented as:

$$\text{Complex}_i : E_{\text{pred},i} \approx E_{\text{exp},i} \tag{3.13}$$

This representation emphasizes the goal of achieving agreement between predicted and experimental binding affinities for each molecular complex.

Validation and benchmarking provide a quantitative assessment of the accuracy of molecular docking methods, guiding researchers in selecting and optimizing algorithms for specific applications.

3.1.8   Limitations of Force Fields

Despite their widespread use, force fields in molecular docking simulations have inherent limitations that researchers must be aware of. Understanding these limitations is crucial for interpreting results and making informed decisions about the reliability of predictions.

Inherent Approximations

Force fields rely on several approximations to simplify the representation of molecular interactions. One significant approximation is the use of fixed parameters for bond lengths, angles, and torsions. This rigid treatment may not fully capture the dynamic nature of molecular structures.

Treatment of Solvation Effects

Force fields often neglect explicit solvent molecules in docking simulations, relying on implicit solvent models. While these models are computationally efficient, they may not accurately represent the effects of solvent molecules on ligand-receptor interactions, especially in cases where solvation plays a crucial role.

Accuracy of Energy Minimization

The accuracy of molecular docking heavily depends on the effectiveness of energy minimization algorithms. However, energy minimization may converge to local minima rather than the global minimum, impacting the accuracy of predicted binding conformations.

Parameterization Challenges

Parameterization of force fields involves assigning values to various parameters, such as Van der Waals radii and partial charges. Obtaining accurate parameter values for diverse molecular structures is a challenging task, and inaccuracies in parameterization can lead to errors in predicted binding energies.

Chemical Formulas

The limitations of force fields can be represented in chemical formulas, emphasizing the challenges associated with approximations and parameterization:

$$
\begin{aligned}
\text{Force Field Limitation} : {} & \text{Inherent Approximations} \\
& + \text{Treatment of Solvation Effects} \\
& + \text{Accuracy of Energy Minimization} \\
& + \text{Parameterization Challenges}
\end{aligned}
\tag{3.14}
$$

Acknowledging these limitations is essential for researchers to critically evaluate the reliability and applicability of molecular docking results. It also underscores the ongoing efforts to improve force fields and develop more accurate computational models.

3.1.9 Force Fields in Molecular Docking Simulations

Force fields play a pivotal role in molecular docking simulations, providing the mathematical framework to calculate the potential energy of molecular complexes. These force fields incorporate various terms to model different types of interactions between atoms, ensuring an accurate representation of the binding interactions.

Total Potential Energy Calculation

The total potential energy (E_{total}) of a molecular complex in a force field can be calculated as the sum of various contributions:

$$
E_{\text{total}} = E_{\text{bonded}} + E_{\text{non-bonded}}
\tag{3.15}
$$

Here, E_{bonded} represents the energy associated with bonded interactions, including bonds, angles, and torsions, while $E_{\text{non-bonded}}$ accounts for non-bonded interactions, such as Van der Waals forces and electrostatic interactions.

Bonded Interactions

The energy associated with bonded interactions can be expressed as:

$$E_{\text{bonded}} = E_{\text{bonds}} + E_{\text{angles}} + E_{\text{torsions}} \tag{3.16}$$

Where:

- E_{bonds} represents the energy associated with bond stretching.

- E_{angles} accounts for the energy associated with angle bending.

- E_{torsions} captures the energy related to torsional rotations.

Non-Bonded Interactions

The energy associated with non-bonded interactions is typically divided into Van der Waals (E_{vdW}) and electrostatic (E_{elec}) contributions:

$$E_{\text{non-bonded}} = E_{\text{vdW}} + E_{\text{elec}} \tag{3.17}$$

Here:

- E_{vdW} represents the Van der Waals interaction energy.

- E_{elec} accounts for the electrostatic interaction energy.

Chemical Formulas

The representation of force fields in molecular docking simulations can be expressed as:

$$
\begin{aligned}
\text{Force Field Model}: E_{\text{total}} \\
= E_{\text{bonded}} + E_{\text{non-bonded}} \\
= E_{\text{bonds}} + E_{\text{angles}} + E_{\text{torsions}} + E_{\text{vdW}} + E_{\text{elec}}
\end{aligned}
\tag{3.18}
$$

Understanding the contributions of bonded and non-bonded interactions is essential for interpreting the energetics of molecular docking simulations and making informed decisions about ligand-receptor binding.

3.1.10 Advanced Force Field Techniques

As molecular docking simulations continue to evolve, researchers have developed advanced force field techniques to address specific challenges and enhance the accuracy of predictions. These techniques often incorporate sophisticated mathematical formulations and chemical considerations.

Enhanced Electrostatic Models

One advanced technique involves refining the electrostatic models used in force fields. The Coulombic interaction (E_{elec}) is often enhanced by incorporating polarizability effects:

$$E_{\text{elec}} = \frac{q_i \cdot q_j}{\varepsilon r_{ij}} + \alpha_{ij} \cdot \frac{(q_i \cdot q_j)^2}{\varepsilon r_{ij}^3} \tag{3.19}$$

Here, α_{ij} represents the polarizability term, accounting for the ability of molecules to respond to electric fields.

Inclusion of Quantum Mechanical Corrections

To improve the accuracy of force fields, quantum mechanical corrections can be included for specific regions of interest. This hybrid approach, known as quantum mechanics/molecular mechanics (QM/MM), combines the accuracy of quantum mechanical calculations with the efficiency of molecular mechanics:

$$E_{\text{QM/MM}} = E_{\text{QM}} + E_{\text{MM}} + E_{\text{QM/MM}} \tag{3.20}$$

Here, E_{QM} represents the quantum mechanical energy, E_{MM} is the molecular mechanics energy, and $E_{\text{QM/MM}}$ captures the interactions between the quantum and classical regions.

Chemical Formulas

The advanced force field techniques can be summarized in chemical formulas:

Advanced Force Field Technique : E_{elec}

$$= \frac{q_i \cdot q_j}{\varepsilon r_{ij}} + \alpha_{ij} \cdot \frac{(q_i \cdot q_j)^2}{\varepsilon r_{ij}^3} + E_{\text{QM/MM}} \quad (3.21)$$

$$= E_{\text{QM}} + E_{\text{MM}} + E_{\text{QM/MM}}$$

These advancements highlight the ongoing efforts to refine force field models, enabling more accurate predictions of ligand-receptor interactions in molecular docking simulations.

3.1.11   Case Studies: Force Field Applications

In this section, we explore real-world case studies that demonstrate the practical applications of force fields in molecular docking simulations. These case studies showcase the effectiveness of force field models in predicting and analyzing ligand-receptor interactions.

Case Study 1: Protein-Ligand Binding Affinity Prediction

One common application of force fields is predicting the binding affinity (ΔG_{bind}) between a protein and a ligand. The binding affinity can be estimated using the following formula:

$$\Delta G_{\text{bind}} = \Delta E_{\text{total}} - T\Delta S \quad (3.22)$$

Here, ΔE_{total} is the total energy change upon binding, T is the temperature, and ΔS represents the entropy change.

Case Study 2: Ligand Conformational Analysis

Force fields are instrumental in studying the conformational changes of ligands upon binding. The conformational energy (E_{conf}) can be calculated as the sum of bond, angle, and torsion contributions:

$$E_{\text{conf}} = E_{\text{bonds}} + E_{\text{angles}} + E_{\text{torsions}} \quad (3.23)$$

This analysis aids in understanding the energetics of ligand flexibility.

Chemical Formulas

The application of force fields in case studies can be expressed in chemical formulas:

$$\text{Case Study 1}: \Delta G_{\text{bind}} = \Delta E_{\text{total}} - T\Delta S \tag{3.24}$$

$$\text{Case Study 2}: E_{\text{conf}} = E_{\text{bonds}} + E_{\text{angles}} + E_{\text{torsions}} \tag{3.25}$$

These case studies exemplify the versatility of force fields in addressing diverse aspects of molecular docking, providing valuable insights for drug discovery and structural biology.

3.1.12   Machine Learning and Force Fields

The integration of machine learning techniques with traditional force fields has emerged as a powerful approach to enhance the accuracy and efficiency of molecular docking simulations. This synergy leverages the strengths of both methodologies to address complex challenges in predicting ligand-receptor interactions.

Hybrid Potential Energy Functions

Machine learning models are often employed to refine potential energy functions in force fields. The hybrid potential energy (E_{hybrid}) can be formulated as a combination of traditional force field terms (E_{FF}) and machine learning corrections (E_{ML}):

$$E_{\text{hybrid}} = w \cdot E_{\text{FF}} + (1 - w) \cdot E_{\text{ML}} \tag{3.26}$$

Here, w represents the weight assigned to the traditional force field.

Data-Driven Parameterization

Machine learning algorithms are employed to optimize force field parameters based on experimental or quantum mechanical data. The parameterization process can be expressed as an optimization problem:

$$\text{Minimize} \sum_i (E_{\text{FF}}^i - E_{\text{exp/qm}}^i)^2 \tag{3.27}$$

This minimization aims to adjust force field parameters (θ) to minimize the difference between force field-calculated energies (E_{FF}^i) and reference energies ($E_{\text{exp/qm}}^i$).

Chemical Formulas

The integration of machine learning with force fields can be summarized in chemical formulas:

$$\text{Hybrid Potential Energy}: E_{\text{hybrid}} = w \cdot E_{\text{FF}} + (1 - w) \cdot E_{\text{ML}} \tag{3.28}$$

$$\text{Parameterization}: \text{Minimize} \sum_i (E_{\text{FF}}^i - E_{\text{exp/qm}}^i)^2 \tag{3.29}$$

These approaches showcase the potential of combining machine learning and force fields for more accurate and adaptable molecular docking simulations.

3.1.13   Future Directions in Force Field Development

The future of force fields holds exciting prospects. Ongoing research focuses on refining parameters, improving accuracy in capturing quantum effects, and developing force fields tailored for specific molecular systems. Collaboration between computational chemists, physicists, and machine learning experts is key to advancing force field methodologies.

3.2 Search Algorithms

Search algorithms are the backbone of molecular docking simulations, responsible for exploring the vast conformational space of ligands within the receptor binding site. In this section, we delve into the various search algorithms used in molecular docking, their principles, and their applications.

3.2.1 Importance of Search Algorithms

Efficient search algorithms play a crucial role in molecular docking simulations by exploring the vast conformational space to identify energetically favorable ligand-receptor poses. Understanding the significance of search algorithms is essential for optimizing the accuracy and speed of docking studies.

Scoring Function Optimization

Search algorithms are often employed to optimize the scoring functions used to evaluate ligand-receptor interactions. The scoring function (S) can be refined through iterative optimization:

$$S_{\text{optimized}} = S_{\text{initial}} + \Delta S \tag{3.30}$$

Here, ΔS represents the improvement achieved through the search algorithm.

Global vs. Local Search Strategies

Different search algorithms employ either global or local search strategies. A global search explores a broader conformational space, while a local search focuses on refining the conformation around an initial pose. The balance between global and local strategies impacts the thoroughness of the search.

Chemical Formulas

The importance of search algorithms in molecular docking can be expressed in chemical formulas:

$$\text{Scoring Function Optimization} : S_{\text{optimized}} = S_{\text{initial}} + \Delta S \tag{3.31}$$

These considerations underscore the critical role of search algorithms in enhancing the accuracy and efficiency of molecular docking simulations, ultimately contributing to successful ligand binding predictions.

3.2.2 Exhaustive Search Methods

Exhaustive search methods aim to systematically explore the entire conformational space of ligand-receptor interactions, providing a comprehensive analysis of potential binding poses. These methods are characterized by their thoroughness but can be computationally demanding.

Grid-Based Search

In a grid-based search, the conformational space is discretized into a three-dimensional grid. The ligand is then systematically placed at each grid point, and the scoring function is evaluated. The mathematical representation can be defined as:

$$S_{\text{grid}}(x, y, z) = \sum_i w_i \cdot \phi_i(x, y, z) \tag{3.32}$$

Here, x, y, and z represent the grid coordinates, w_i are weights, and ϕ_i are scoring function terms.

Brute-Force Enumeration

Brute-force enumeration involves exhaustively sampling all possible ligand conformations within predefined ranges. The mathematical representation can be expressed as:

$$S_{\text{brute-force}} = \sum_j w_j \cdot \psi_j \tag{3.33}$$

Here, w_j are weights, and ψ_j are scoring function terms for each sampled conformation.

Chemical Formulas

The exhaustive search methods can be summarized in chemical formulas:

$$\text{Grid-Based Search} : S_{\text{grid}}(x, y, z) = \sum_i w_i \cdot \phi_i(x, y, z) \tag{3.34}$$

$$\text{Brute-Force Enumeration} : S_{\text{brute-force}} = \sum_j w_j \cdot \psi_j \tag{3.35}$$

These methods showcase the meticulous exploration of conformational space, offering a detailed understanding of ligand binding interactions.

3.2.3 Monte Carlo Methods

Monte Carlo methods provide a probabilistic approach to exploring ligand-receptor conformational space, emphasizing stochastic sampling. These methods leverage random sampling to uncover energetically favorable binding poses.

Metropolis Algorithm

The Metropolis algorithm is a classic Monte Carlo method used in molecular docking. It involves proposing a new ligand conformation and accepting or rejecting it based on the Metropolis criterion:

$$P_{\text{accept}} = \begin{cases} e^{-\Delta E / kT}, & \text{if } \Delta E > 0 \\ 1, & \text{if } \Delta E \leq 0 \end{cases} \tag{3.36}$$

Here, ΔE is the energy difference between the current and proposed conformations, k is the Boltzmann constant, and T is the temperature.

Chemical Formulas

The Monte Carlo methods can be expressed in chemical formulas:

$$\text{Metropolis Algorithm}: P_{\text{accept}} = \begin{cases} e^{-\Delta E/kT}, & \text{if } \Delta E > 0 \\ 1, & \text{if } \Delta E \leq 0 \end{cases} \tag{3.37}$$

These methods provide a powerful means to explore ligand-receptor interactions, introducing stochasticity to the sampling process.

3.2.4   Genetic Algorithms

Genetic algorithms are optimization techniques inspired by natural selection and genetics. In the context of molecular docking, they are employed to search for optimal ligand conformations and binding poses.

Genetic Operators

Genetic algorithms typically involve three main operators: selection, crossover, and mutation.

1. **Selection:** Individuals (ligand conformations) are selected for reproduction based on their fitness, which is determined by the scoring function.

2. **Crossover:** Pairs of selected individuals exchange genetic information to generate new conformations. The crossover operation can be represented as:

$$\text{Crossover}(A, B) = \frac{A + B}{2} \tag{3.38}$$

Here, A and B represent two parent conformations.

3. **Mutation:** Random changes are introduced to individual conformations to explore new regions of the conformational space. The mutation operation can be expressed as:

$$\text{Mutation}(C) = C + \Delta C \tag{3.39}$$

Here, C is an individual conformation, and ΔC represents a random perturbation.

Chemical Formulas

The genetic algorithms can be summarized in chemical formulas:

$$\text{Crossover Operation} : \text{Crossover}(A, B) = \frac{A + B}{2} \tag{3.40}$$

$$\text{Mutation Operation} : \text{Mutation}(C) = C + \Delta C \tag{3.41}$$

These algorithms emulate evolutionary processes to efficiently explore ligand-receptor conformational space.

3.2.5   Lamarckian Genetic Algorithms

Lamarckian genetic algorithms introduce a mechanism for the direct adaptation of individuals' genetic information based on the success of their offspring during the search process.

Adaptation Mechanism

The adaptation mechanism in Lamarckian genetic algorithms involves modifying the genetic information of individuals based on the fitness of their offspring. This can be expressed as:

$$\text{Adaptation}(A, B) = A + \alpha \cdot (B - A) \tag{3.42}$$

Here, A represents the genetic information of the parent, B is the genetic information of the offspring, and α is a parameter determining the extent of adaptation.

Chemical Formulas

The Lamarckian genetic algorithms can be represented in chemical formulas:

$$\text{Adaptation Mechanism}: \text{Adaptation}(A, B) = A + \alpha \cdot (B - A) \qquad (3.43)$$

These algorithms combine the principles of genetic algorithms with a Lamarckian inheritance strategy, facilitating a more direct adaptation of individuals to their environment.

3.2.6   Ant Colony Optimization

Ant Colony Optimization (ACO) is a metaheuristic inspired by the foraging behavior of ants. It is commonly applied to combinatorial optimization problems, including molecular docking.

Pheromone Update Rule

The central idea of ACO is the use of pheromones to communicate information about the quality of solutions. The pheromone update rule is a key component and can be expressed as:

$$\tau_{ij} = (1 - \rho) \cdot \tau_{ij} + \Delta\tau_{ij} \qquad (3.44)$$

Here, τ_{ij} represents the pheromone level on the edge between positions i and j, ρ is the pheromone evaporation rate, and $\Delta\tau_{ij}$ is the pheromone deposit by ants.

Chemical Formulas

The pheromone update rule in chemical formulas:

$$\text{Pheromone Update Rule}: \tau_{ij} = (1 - \rho) \cdot \tau_{ij} + \Delta\tau_{ij} \qquad (3.45)$$

ACO leverages the collective intelligence of ants to efficiently explore ligand-receptor conformational space, making it a valuable optimization technique.

3.2.7 Hybrid Search Strategies

Hybrid search strategies are designed to combine the strengths of multiple optimization techniques, overcoming individual limitations and enhancing overall search efficiency in molecular docking studies.

Combination of Genetic Algorithms and Local Search

One effective hybrid strategy involves combining Genetic Algorithms (GA) for global exploration with Local Search (LS) methods for local refinement. This combination can be represented as:

$$\text{Hybrid Strategy}: \text{Combine}(GA, LS) = \alpha \cdot \text{GA} + (1 - \alpha) \cdot \text{LS} \qquad (3.46)$$

Here, α is a parameter that controls the balance between global and local search. Adjusting α allows researchers to tailor the hybrid strategy based on the specific characteristics of the docking problem.

Chemical Formulas

In chemical formulas, the hybrid strategy combining genetic algorithms and local search is represented as:

$$\text{Hybrid Strategy}: \text{Combine}(GA, LS) = \alpha \cdot \text{GA} + (1 - \alpha) \cdot \text{LS} \qquad (3.47)$$

This hybrid approach aims to synergistically exploit the exploration capabilities of genetic algorithms and the local refinement capabilities of local search methods, contributing to more effective ligand-receptor conformational sampling in molecular docking simulations.

3.2.8 Adaptive Search Algorithms

Adaptive search algorithms dynamically adjust their behavior during the optimization process based on the evolving landscape of the search space. These algorithms aim to enhance efficiency and convergence by adapting their strategies to the characteristics of the molecular docking problem.

Adaptive Genetic Algorithms (AGA)

Adaptive Genetic Algorithms (AGA) incorporate mechanisms to dynamically adjust parameters such as mutation rates and crossover probabilities during the optimization process. The adaptation can be represented as:

$$\text{Mutation Rate}_{\text{new}} = \text{Adaptation Function}(\text{Mutation Rate}_{\text{old}}) \qquad (3.48)$$

This adaptation allows AGAs to respond to the specific challenges encountered in the ligand-receptor conformational search.

Chemical Formulas

In chemical formulas, the adaptive mutation rate in Adaptive Genetic Algorithms is represented as:

$$\text{Mutation Rate}_{\text{new}} = \text{Adaptation Function}(\text{Mutation Rate}_{\text{old}}) \qquad (3.49)$$

Adaptive search algorithms, such as AGAs, play a crucial role in addressing the dynamic and complex nature of molecular docking problems by adjusting their exploration-exploitation balance during the optimization process.

3.2.9   Parallelization and High-Performance Computing

Utilizing parallelization and high-performance computing (HPC) techniques is crucial for accelerating molecular docking simulations and handling computationally demanding tasks efficiently.

Parallel Genetic Algorithms (PGA)

Parallel Genetic Algorithms (PGA) leverage parallel processing to enhance the exploration of the search space. In parallelization, the fitness evaluations of multiple individuals can be performed simultaneously. The parallelization of genetic algorithms can be expressed as:

$$\text{Parallelization} : \text{PGA} = \text{Parallelize(GA)} \tag{3.50}$$

This parallelization approach significantly reduces the computation time required for large-scale docking studies.

Chemical Formulas

In chemical formulas, the parallelization of genetic algorithms is represented as:

$$\text{Parallelization} : \text{PGA} = \text{Parallelize(GA)} \tag{3.51}$$

High-performance computing techniques, including parallelization, enable researchers to efficiently explore complex ligand-receptor conformational spaces and accelerate the drug discovery process.

3.2.10 Case Studies: Search Algorithm Applications

Exploring real-world case studies helps illustrate the practical applications of various search algorithms in molecular docking scenarios. The following examples highlight the effectiveness of different algorithms in addressing specific challenges.

Case Study 1: Genetic Algorithm in Flexible Docking

In a flexible docking scenario, the Genetic Algorithm (GA) is employed to explore the conformational space of a flexible ligand interacting with a rigid receptor. The objective function considers both binding affinity and flexibility-related terms:

$$\text{Objective Function} = \text{Binding Affinity} - \text{Flexibility Penalty} \tag{3.52}$$

This formulation allows the GA to efficiently search for energetically favorable conformations considering ligand flexibility.

Case Study 2: Monte Carlo Method in Solvent Effects

The Monte Carlo method is applied to study the impact of solvent effects on ligand binding. The free energy of binding in the presence of solvent can be expressed as:

$$\Delta G_{\text{solvent}} = \Delta G_{\text{gas phase}} + \Delta G_{\text{solvation}} \tag{3.53}$$

Here, the Monte Carlo method aids in sampling solvent configurations and estimating solvation free energy contributions.

Chemical Formulas

In chemical formulas, the objective function for Genetic Algorithm in flexible docking is represented as:

$$\text{Objective Function} = \text{Binding Affinity} - \text{Flexibility Penalty} \tag{3.54}$$

3.2.11   Machine Learning in Search Algorithms

Integrating machine learning (ML) techniques with search algorithms enhances the efficiency and accuracy of molecular docking studies. Machine learning models can learn patterns from large datasets, aiding in the optimization of search strategies.

Machine Learning-Assisted Genetic Algorithms

In machine learning-assisted genetic algorithms, a predictive model (MLP) is trained to estimate the fitness of individuals based on their genetic information. The fitness prediction can be formulated as:

$$\text{Fitness Prediction} = \text{MLP}(\text{Genetic Information}) \tag{3.55}$$

This allows the genetic algorithm to prioritize promising individuals, improving convergence speed.

Chemical Formulas

In chemical formulas, the fitness prediction in machine learning-assisted genetic algorithms is represented as:

$$\text{Fitness Prediction} = \text{MLP}(\text{Genetic Information}) \tag{3.56}$$

Machine learning integration optimizes search algorithms, making them adaptable to specific docking scenarios and accelerating the exploration of ligand-receptor conformational spaces.

3.2.12 Validation and Benchmarking

Ensuring the reliability and accuracy of molecular docking results requires thorough validation and benchmarking against known datasets. Various metrics and approaches are employed for this purpose.

Scoring Function Validation

To validate a scoring function's performance, the root mean square deviation (RMSD) between predicted and experimental binding poses can be calculated using the formula:

$$\text{RMSD} = \sqrt{\frac{1}{N} \sum_{i=1}^{N} (\text{Experimental Position} - \text{Predicted Position})^2} \tag{3.57}$$

Here, N represents the number of atoms in the ligand. A lower RMSD indicates better agreement.

Benchmarking Against Experimental Data

Benchmarking involves comparing docking results against experimental binding affinities. The correlation coefficient (R^2) is commonly used and is calculated as:

$$X = \left(\sum_{i=1}^{N} (\text{Experimental Affinity}_i - \overline{\text{Experimental Affinity}})(\text{Predicted Affinity}_i \right.$$

$$\left. - \overline{\text{Predicted Affinity}}) \right)^2$$

$$Y = \left(\sum_{i=1}^{N} (\text{Experimental Affinity}_i \right.$$

$$\left. - \overline{\text{Experimental Affinity}})^2 \right) \left(\sum_{i=1}^{N} (\text{Predicted Affinity}_i \right.$$

$$\left. - \overline{\text{Predicted Affinity}})^2 \right)$$

$$\implies R^2 = \frac{X}{Y}$$

$$(3.58)$$

where N is the number of compounds,

and $\overline{\text{Experimental Affinity}}$ and $\overline{\text{Predicted Affinity}}$ are the mean experimental and predicted affinities, respectively.

Chemical Formulas

In chemical formulas, the RMSD and correlation coefficient formulas are represented as:

$$\text{RMSD} = \sqrt{\frac{1}{N} \sum_{i=1}^{N} (\text{Experimental Position} - \text{Predicted Position})^2} \qquad (3.59)$$

$$X = \left(\sum_{i=1}^{N} (\text{Experimental Affinity}_i - \overline{\text{Experimental Affinity}})(\text{Predicted Affinity}_i \right.$$
$$\left. - \overline{\text{Predicted Affinity}}) \right)^2$$

$$Y$$
$$= \left(\sum_{i=1}^{N} (\text{Experimental Affinity}_i \right.$$
$$\left. - \overline{\text{Experimental Affinity}})^2 \right) \left(\sum_{i=1}^{N} (\text{Predicted Affinity}_i \right.$$
$$\left. - \overline{\text{Predicted Affinity}})^2 \right)$$

$$\implies R^2$$
$$= \frac{X}{Y}$$

$$(3.60)$$

Validation and benchmarking are crucial steps to assess the predictive power of molecular docking methodologies.

3.2.13 Limitations of Search Algorithms

While search algorithms play a crucial role in exploring the conformational space of ligand-receptor interactions, they come with certain limitations that impact their effectiveness.

Local Minima

One common limitation is the susceptibility to getting trapped in local minima during the optimization process. Local minima are suboptimal configurations that the algorithm converges to, preventing the exploration of the global minimum. Mathematically, this can be expressed as:

$$f(\text{Local Minimum}) < f(\text{Global Minimum}) \qquad (3.61)$$

where f represents the objective function being minimized.

Chemical Formulas

The issue of local minima in chemical formulas:

$$f(\text{Local Minimum}) < f(\text{Global Minimum}) \qquad (3.62)$$

Convergence Rate

The convergence rate of search algorithms is another limitation. Some algorithms may converge slowly, requiring a large number of iterations to reach an optimal solution. The convergence rate (CR) is defined as:

$$CR = \frac{\text{Initial Value} - \text{Optimal Value}}{\text{Initial Value}} \qquad (3.63)$$

A slower convergence rate can hinder the practical applicability of the algorithm.

Chemical Formulas

The convergence rate in chemical formulas:

$$CR = \frac{\text{Initial Value} - \text{Optimal Value}}{\text{Initial Value}} \qquad (3.64)$$

Understanding and addressing these limitations are essential for the development and improvement of search algorithms in molecular docking simulations.

3.2.14   Future Directions in Search Algorithm Development

The field of search algorithms in molecular docking is dynamic, and ongoing research is focused on addressing existing challenges and exploring new avenues for improvement. Several promising directions for future development include:

Incorporation of Machine Learning

One potential avenue is the integration of machine learning techniques to enhance the efficiency and accuracy of search algorithms. The incorporation of predictive models can guide the search process and reduce the computational burden. Mathematically, this integration can be represented as:

$$\text{Hybrid Algorithm : Combine}(SA, ML) = \beta \cdot \text{SA} + (1 - \beta) \cdot \text{ML} \qquad (3.65)$$

where SA is the traditional search algorithm, and ML is the machine learning component, controlled by the parameter β.

Chemical Formulas

The chemical formulas representing the integration of machine learning in search algorithms:

$$\text{Hybrid Algorithm : Combine}(SA, ML) = \beta \cdot \text{SA} + (1 - \beta) \cdot \text{ML} \qquad (3.66)$$

Parallelization and Distributed Computing

Another direction involves leveraging parallelization and distributed computing to accelerate search algorithms. Distributing the computational workload across multiple processors or nodes can significantly reduce the overall simulation time. The parallelization strategy can be expressed as:

$$\text{Parallelized Algorithm : Parallelize}(SA) = \gamma \cdot \text{SA} \qquad (3.67)$$

where SA is the original search algorithm, and γ controls the degree of parallelization.

Chemical Formulas

The chemical formulas representing the parallelization of search algorithms:

$$\text{Parallelized Algorithm} : \text{Parallelize}(SA) = \gamma \cdot \text{SA} \qquad (3.68)$$

These future directions hold the potential to significantly advance the field of search algorithms in molecular docking, paving the way for more efficient and accurate ligand-receptor interaction predictions.

Chapter 4

Data Preparation

4.1 Preparation of Ligands

The preparation of ligands is a critical phase in molecular docking studies, ensuring accurate and reliable simulations of ligand-receptor interactions. This process involves several key steps, each designed to enhance the quality of ligand structures and optimize their compatibility with the docking algorithm.

4.1.1 Structure Retrieval

Structure retrieval in molecular docking involves the identification and extraction of relevant molecular conformations based on specific criteria. This process is crucial for obtaining meaningful insights into ligand-receptor interactions.

Mathematical Formulas

One common mathematical formula for structure retrieval based on a scoring function could be expressed as follows:

$$\text{Score}(C) = \sum_{i=1}^{N} w_i \cdot f_i(C) \tag{4.1}$$

Here, C represents a molecular conformation, N is the number of scoring terms, w_i is the weight assigned to the i-th term, and $f_i(C)$ is the value of the i-th scoring term for the conformation C.

Chemical Formulas

The chemical formulas involved in structure retrieval may include specific descriptors or properties used for filtering conformations. For instance, if a certain bond length or angle is a critical criterion, it could be represented as:

$$\text{Bond Length Criterion}: \quad d_{ij} = \sqrt{(x_i - x_j)^2 + (y_i - y_j)^2 + (z_i - z_j)^2} \quad (4.2)$$

Where d_{ij} is the bond length between atoms i and j, and (x_i, y_i, z_i) are the coordinates of atom i.

Structure retrieval algorithms often combine mathematical and chemical criteria to efficiently identify conformations that meet specific requirements, contributing to the accuracy and relevance of docking results.

4.1.2   Geometry Optimization

Geometry optimization in molecular docking aims to find the most energetically favorable conformation by adjusting the spatial arrangement of atoms. This process involves minimizing the energy of the molecular system with respect to its coordinates.

Mathematical Formulas

A common mathematical formulation for energy minimization can be expressed using a potential energy function E:

$$\min_{\text{coordinates}} E(\text{coordinates}) \qquad (4.3)$$

Here, coordinates represent the spatial positions of atoms, and the minimization is performed to find the optimal set of coordinates that minimizes the potential energy.

Chemical Formulas

The chemical aspect of geometry optimization involves adjusting bond lengths, bond angles, and torsional angles to achieve stable and realistic molecular conformations. Specific chemical formulas for adjusting bond lengths or angles may include:

$$\text{Adjusting Bond Lengths}: \quad d_{ij} = \sqrt{(x_i - x_j)^2 + (y_i - y_j)^2 + (z_i - z_j)^2} \quad (4.4)$$

Where d_{ij} is the bond length between atoms i and j, and (x_i, y_i, z_i) are the coordinates of atom i.

Geometry optimization plays a crucial role in refining the predicted ligand-receptor complexes, ensuring that the obtained conformations are energetically favorable and biologically relevant.

4.1.3 Tautomer and Stereoisomer Handling

In molecular docking, considering tautomers and stereoisomers is crucial for accurately representing ligand structures and exploring their potential binding modes. Tautomers are isomers that can be interconverted by the migration of a proton, while stereoisomers have the same molecular formula and connectivity but differ in the spatial arrangement of atoms.

Tautomer Handling

Handling tautomers involves considering all possible protonation states and interconversions. A mathematical representation of tautomer handling can be expressed as:

$$\begin{aligned}
\text{Tautomer Enumeration} \\
: \quad \text{Ligand} \\
\rightleftharpoons \text{Tautomer}_1 \\
\rightleftharpoons \text{Tautomer}_2 \\
\rightleftharpoons \ldots
\end{aligned} \qquad (4.5)$$

This equation represents the dynamic equilibrium between different tautomeric forms of a ligand during the docking process.

Stereoisomer Handling

Stereoisomer handling requires considering different spatial arrangements of atoms without changing the molecular formula. This can be represented as:

$$\text{Stereoisomer Enumeration}$$
$$: \quad \text{Ligand} \xrightleftharpoons[\text{Stereoisomerization}]{\text{Rotation around Stereo Bonds}} \text{Stereoisomer}_1 \qquad (4.6)$$
$$\xrightleftharpoons[\text{Stereoisomerization}]{\text{Rotation around Stereo Bonds}} \text{Stereoisomer}_2 \xrightleftharpoons[\cdots]{\cdots}$$

This equation illustrates the exploration of different stereoisomeric forms of a ligand, considering rotations around stereocenters.

Tautomer and stereoisomer handling ensures a comprehensive exploration of ligand conformational space, improving the accuracy of molecular docking predictions.

4.1.4 Ionization State Determination

In molecular docking, considering the ionization state of ligands is essential as it influences their binding affinity and interactions with receptors. The ionization state determination involves predicting the protonation and deprotonation states of ionizable groups within a ligand.

Mathematical Representation

A mathematical representation of ionization state determination can be expressed as:

$$\text{Ionization State Determination}$$
$$: \quad \text{Ligand} \xrightleftharpoons[\text{Protonation}]{\text{Deprotonation}} \text{Ionization State}_1 \xrightleftharpoons[\text{Protonation}]{\text{Deprotonation}} \text{Ionization State}_2 \xrightleftharpoons[\cdots]{\cdots}$$

$$(4.7)$$

This equation represents the dynamic equilibrium between different ionization states of a ligand, where protonation and deprotonation processes are considered.

Chemical Formula

Chemically, the ionization state determination can be illustrated using the example of a carboxylic acid group:

$$\text{Ionization State Determination}: \quad COOH \rightleftharpoons COO^- + H^+ \tag{4.8}$$

This equation shows the ionization of a carboxylic acid group into its deprotonated (COO-) and protonated (H+) forms.

Considering ionization states is crucial for accurately modeling ligand-receptor interactions under different physiological conditions.

4.1.5 Parameterization for Molecular Mechanics

Parameterization in molecular mechanics involves assigning numerical values to force field parameters, such as bond lengths, bond angles, dihedral angles, and van der Waals parameters, to accurately represent molecular interactions.

Mathematical Representation

The mathematical representation of parameterization for a bond length (r) can be expressed as:

$$V_{\text{bond}}(r) = \frac{1}{2}k_{\text{bond}}(r - r_{\text{eq}})^2 \tag{4.9}$$

Here, k_{bond} is the force constant, r is the instantaneous bond length, and r_{eq} is the equilibrium bond length.

Chemical Formula

For a chemical example, consider the bond stretching in a diatomic molecule:

$$V_{\text{bond}}(r) = \frac{1}{2} k_{\text{bond}} (r - r_{\text{eq}})^2 \tag{4.10}$$

This formula represents the potential energy associated with stretching a bond, where k_{bond} and r_{eq} are parameters specific to the bond.

Accurate parameterization ensures that molecular mechanics simulations reflect experimental observations and provide reliable insights into molecular behavior.

4.1.6   Adding Hydrogen Atoms

The addition of hydrogen atoms to molecular structures is a crucial step in preparing ligands and receptors for molecular docking simulations. Hydrogen atoms play a significant role in defining the geometry and energetics of molecular interactions.

Mathematical Representation

The mathematical representation of adding hydrogen atoms involves determining the optimal positions and orientations based on the molecular structure. This can be expressed as:

$$E_{\text{addH}} = \sum_i w_i \cdot E_i \tag{4.11}$$

Here, E_{addH} is the total energy associated with adding hydrogen atoms, E_i represents individual energy terms, and w_i are weights assigned to each term.

Chemical Formula

In a chemical formula, the addition of hydrogen atoms can be symbolized as:

$$E_{\text{addH}} = \sum_i w_i \cdot E_i \tag{4.12}$$

This process ensures that the molecular structure is correctly saturated with hydrogen atoms, reflecting the expected protonation states and optimizing the system for docking studies.

Adding hydrogen atoms is a critical step in generating accurate and realistic molecular structures for docking simulations.

4.1.7 Flexible Ligand Considerations

In molecular docking, accounting for ligand flexibility is crucial for capturing the dynamic nature of ligand-receptor interactions. Flexible ligands can adopt different conformations, allowing a more comprehensive exploration of binding modes.

Mathematical Representation

The mathematical representation of ligand flexibility involves considering multiple conformations during the docking process. This can be expressed as a weighted sum of energies for each conformation:

$$E_{\text{flex}} = \sum_j p_j \cdot E_j \tag{4.13}$$

Here, E_{flex} is the total energy associated with flexible ligand considerations, E_j represents the energy of the j-th conformation, and p_j are the probabilities associated with each conformation.

Chemical Formula

In a chemical formula, the flexible ligand considerations can be represented as:

$$E_{\text{flex}} = \sum_j p_j \cdot E_j \tag{4.14}$$

Integrating ligand flexibility enhances the accuracy of docking simulations by accounting for the variability in ligand structures, resulting in more reliable binding predictions.

4.1.8 Protonation State and Tautomer Sensitivity

The protonation state and tautomer of a molecule significantly influence its interactions in molecular docking studies. Accurate determination of these states is crucial for reliable predictions.

Mathematical Representation

The mathematical representation of protonation state and tautomer sensitivity involves considering various possible states and their associated energies. This can be expressed as:

$$E_{\text{prot-taut}} = \min_i \left(E_{\text{prot}_i} + E_{\text{taut}_i} \right) \tag{4.15}$$

Here, $E_{\text{prot-taut}}$ is the total energy considering protonation and tautomer sensitivity, E_{prot_i} is the energy of the i-th protonation state, and E_{taut_i} is the energy of the i-th tautomer.

Chemical Formula

In a chemical formula, the protonation state and tautomer sensitivity can be represented as:

$$E_{\text{prot-taut}} = \min_i \left(E_{\text{prot}_i} + E_{\text{taut}_i} \right) \tag{4.16}$$

Accurate handling of protonation and tautomer states enhances the precision of molecular docking simulations, providing more realistic insights into ligand-receptor interactions.

4.1.9 Ligand Covalent Modifications

Covalent modifications of ligands play a crucial role in expanding the chemical space that can be explored in molecular docking studies. These modifications involve the formation of new covalent bonds, influencing the ligand's conformation and interactions.

Mathematical Representation

The mathematical representation of ligand covalent modifications can be expressed as a set of transformations applied to the ligand structure:

$$E_{\text{covalent-mod}} = \sum_i (E_{\text{mod}_i}) \tag{4.17}$$

Here, $E_{\text{covalent-mod}}$ represents the total energy considering covalent modifications, and E_{mod_i} is the energy contribution of the i-th covalent modification.

Chemical Formula

In a chemical formula, ligand covalent modifications can be represented as:

$$E_{\text{covalent-mod}} = \sum_i (E_{\text{mod}_i}) \tag{4.18}$$

Incorporating covalent modifications into molecular docking simulations allows the exploration of a broader range of chemical modifications, aiding in the discovery of novel ligands with enhanced binding affinities.

4.1.10 Charge Neutralization

Charge neutralization is a critical step in preparing ligands for molecular docking, especially when dealing with charged molecules. Neutralizing the charges ensures that the ligands have a balanced overall charge, which is essential for accurate interaction predictions.

Mathematical Representation

Mathematically, the process of charge neutralization can be represented as:

$$Q_{\text{neutralized}} = Q_{\text{original}} - Q_{\text{ionizable groups}} \tag{4.19}$$

Here, $Q_{\text{neutralized}}$ is the neutralized charge, Q_{original} is the original charge of the ligand, and $Q_{\text{ionizable groups}}$ is the charge contributed by ionizable groups.

Chemical Formula

In a chemical formula, charge neutralization can be depicted as:

$$Q_{\text{neutralized}} = Q_{\text{original}} - Q_{\text{ionizable groups}} \tag{4.20}$$

Charge neutralization is a crucial step to ensure that the ligands are in a physiologically relevant state for accurate docking simulations.

4.1.11   Validation of Prepared Ligands

Validating the prepared ligands is a crucial step to ensure the accuracy and reliability of molecular docking simulations. Various metrics and checks can be employed to assess the quality of ligand preparation.

Mathematical Validation Metrics

Mathematically, the validation of prepared ligands can be expressed using different metrics. One common metric is the root mean square deviation (RMSD) between the experimental and predicted ligand conformations:

$$RMSD = \sqrt{\frac{1}{N} \sum_{i=1}^{N} (\text{Experimental Position}_i - \text{Predicted Position}_i)^2} \tag{4.21}$$

Here, N is the number of atoms in the ligand, and Experimental Position$_i$ and Predicted Position$_i$ are the coordinates of the i-th atom.

Chemical Formula

In a chemical formula, the validation process can be represented as:

$$RMSD = \sqrt{\frac{1}{N} \sum_{i=1}^{N} (\text{Experimental Position}_i - \text{Predicted Position}_i)^2} \tag{4.22}$$

Validation metrics help ensure that the ligands are prepared in a way that preserves their native conformations, increasing the reliability of subsequent docking simulations.

4.1.12 Case Studies: Ligand Preparation in Drug Discovery

Ligand preparation is a critical step in the drug discovery process, and several case studies highlight its impact on molecular docking outcomes.

Energy Minimization Formula

One aspect involves energy minimization during ligand preparation. The potential energy (E_{total}) can be expressed as a sum of various components:

$$E_{\text{total}} = E_{\text{bonded}} + E_{\text{non-bonded}} + E_{\text{electrostatic}} + E_{\text{vdW}} + E_{\text{solvent}} \quad (4.23)$$

Here, E_{bonded}, $E_{\text{non-bonded}}$, $E_{\text{electrostatic}}$, E_{vdW}, and E_{solvent} represent contributions from bonded interactions, non-bonded interactions, electrostatic forces, van der Waals forces, and solvent effects, respectively.

Chemical Formulas

In a chemical formula, the ligand preparation process can be summarized as:

$$E_{\text{total}} = E_{\text{bonded}} + E_{\text{non-bonded}} + E_{\text{electrostatic}} + E_{\text{vdW}} + E_{\text{solvent}} \quad (4.24)$$

These case studies underscore the importance of meticulous ligand preparation for accurate molecular docking simulations in drug discovery.

4.1.13 Machine Learning Applications in Ligand Preparation

Machine learning (ML) plays a pivotal role in optimizing ligand preparation protocols, enhancing efficiency and accuracy. Several ML-based approaches have been employed in ligand preparation workflows.

Generalized Formula

A generalized formula for incorporating machine learning in ligand preparation can be expressed as:

$$\text{ML\_Score} = w_1 \cdot \text{Feature}_1 + w_2 \cdot \text{Feature}_2 + \ldots + w_n \cdot \text{Feature}_n \quad (4.25)$$

Here, ML_Score represents the machine learning score, and Feature_1, Feature_2, ..., Feature_n are various features derived from ligand properties. $w_1, w_2, \ldots, w_n$ denote the corresponding weights assigned during the ML training process.

Chemical Formulas

In a chemical formula, the integration of machine learning in ligand preparation can be summarized as:

$$\text{ML\_Score} = w_1 \cdot \text{Feature}_1 + w_2 \cdot \text{Feature}_2 + \ldots + w_n \cdot \text{Feature}_n \quad (4.26)$$

Machine learning applications bring a data-driven perspective to ligand preparation, enabling tailored and optimized processes for molecular docking studies.

4.1.14 Challenges and Considerations in Ligand Preparation

Ligand preparation is a critical step in molecular docking, and it comes with its set of challenges and considerations that need to be addressed for reliable results.

Equilibrium of Tautomers

The equilibrium between different tautomeric forms poses a challenge in ligand preparation. The equilibrium equation is given by:

$$\text{Tautomerization:} \quad \text{Ligand} \underset{\text{k}-1}{\overset{\text{k1}}{\rightleftharpoons}} \text{Tautomer} \quad (4.27)$$

Here, k_1 and k_{-1} represent the forward and reverse rate constants, respectively.

Stereoisomer Handling

Stereoisomers, especially those with multiple chiral centers, require careful handling. The enumeration of stereoisomers can be expressed as:

$$\text{Stereoisomers:} \quad \text{Ligand} \rightleftharpoons \text{Isomer}_1 \rightleftharpoons \text{Isomer}_2 \rightleftharpoons \ldots \rightleftharpoons \text{Isomer}_n$$

$$(4.28)$$

The equilibrium involves interconversion between different stereoisomeric forms.

Chemical Formula

Considering these challenges, the overall process can be summarized as:

$$\text{Ligand Preparation Challenges : Tautomerization, Stereoisomer Handling, etc.}$$

$$(4.29)$$

Addressing these challenges is crucial for accurate ligand structures in molecular docking simulations.

4.1.15 Integration with Receptor Preparation

The preparation of ligands is intricately linked with the preparation of receptors. Coordinated efforts in optimizing ligand and receptor structures ensure a realistic representation of molecular interactions. Integration of ligand and receptor preparation is vital for the success of molecular docking studies.

4.1.16 Future Directions in Ligand Preparation

The future of ligand preparation involves continuous refinement of methods and increased integration with experimental data. Advances in quantum mechanics-based approaches, machine learning models, and high-throughput ligand screening techniques are expected to shape the landscape of ligand preparation in the coming years.

4.2   Preparation of Receptors

The preparation of receptors is a crucial aspect of molecular docking, influencing the accuracy and reliability of simulations. This section explores the key steps involved in preparing receptors for docking studies, emphasizing the importance of realistic representations and optimized structures.

4.2.1   Protein Retrieval and Structure Selection

In molecular docking, the proper selection of protein structures is crucial for accurate simulations. The process involves retrieving relevant protein structures from databases and applying criteria for structure selection.

Mathematical Formula for Structure Selection Criteria

One common mathematical formula for structure selection criteria might involve scoring different protein conformations:

$$\text{Score} = w_1 \cdot \text{Scoring\_Term}_1 + w_2 \cdot \text{Scoring\_Term}_2 + \ldots + w_n \cdot \text{Scoring\_Term}_n \quad (4.30)$$

Here, $w_1, w_2, \ldots, w_n$ are weights assigned to different scoring terms.

Chemical Formula for Protein Retrieval

The retrieval of proteins from a database represented in a chemical formula:

$$\text{Protein} \rightarrow \text{Database[Criteria]} \quad (4.31)$$

This indicates the process of selecting proteins from a database based on specific criteria.

Proper protein retrieval and structure selection contribute to the accuracy and reliability of molecular docking simulations.

4.2.2 Removal of Water Molecules and Heteroatoms

In the preparation of ligands for molecular docking simulations, the removal of water molecules and unwanted heteroatoms is a crucial step. This process ensures that the ligands are optimized for accurate interactions with receptors.

Mathematical Formula for Removal Criteria

A mathematical formula representing the criteria for removing water molecules and heteroatoms might involve a scoring function:

$$\text{Score\_Removal} = w_1 \cdot \text{Water\_Score} + w_2 \cdot \text{Heteroatom\_Score} \qquad (4.32)$$

Here, w_1 and w_2 are weights assigned to the respective scoring terms.

Chemical Formula for Removal Process

The removal of water molecules and heteroatoms represented in a chemical formula:

$$\text{Ligand} \xrightarrow{\text{Removal}} \text{Ligand}_{\text{clean}} \qquad (4.33)$$

This indicates the transformation of the ligand by removing water molecules and unwanted heteroatoms.

Proper removal of water molecules and heteroatoms enhances the ligand's suitability for molecular docking simulations.

4.2.3 Addition of Hydrogen Atoms

Adding hydrogen atoms to the protein structure is crucial for a realistic representation of molecular interactions. Tools like CHARMM or PDB2PQR can predict the optimal placement of hydrogen atoms based on the protein's molecular structure and the surrounding environment.

4.2.4   Protonation State Determination

The protonation state of amino acid residues in the protein is pH-dependent. Tools like PROPKA or H++ can predict the protonation states of titratable residues at different pH values, providing insights into the ionization states relevant to physiological conditions.

4.2.5   Energy Minimization

Energy minimization is employed to relax the protein structure and alleviate steric clashes or irregularities introduced during the preparation steps. Force field-based methods, such as steepest descent or conjugate gradient, can be applied to achieve a stable and energetically favorable receptor conformation.

4.2.6   Handling of Missing Side Chains

Some protein structures may have missing side-chain atoms, especially in regions distant from the active site. Predicting or modeling these missing side chains using tools like MODELLER or SCWRL can enhance the completeness of the receptor structure.

4.2.7   Ligand Binding Site Identification

Accurate identification of the ligand binding site is crucial for successful docking simulations. Computational tools, including SiteMap or CASTp, can analyze the protein structure and predict potential binding sites based on factors like solvent accessibility and geometric features.

4.2.8   Flexible Residue Considerations

Certain residues in the binding site may exhibit flexibility and undergo conformational changes upon ligand binding. Molecular dynamics simulations or normal mode analysis can be employed to explore the flexibility of these residues, providing insights into dynamic aspects of ligand-receptor interactions.

4.2.9   Co-factor and Metal Ion Handling

If the protein requires co-factors or metal ions for proper function, these should be retained in the structure. Tools like MetalPDB or CheckMyMetal can assist in validating and handling metal ions, ensuring their correct representation in the receptor structure.

4.2.10   Validation of Prepared Receptors

Validation of prepared receptors involves assessing the quality and accuracy of the structural modifications. Techniques such as Ramachandran plot analysis, MolProbity scores, and comparison with experimental data contribute to the validation process, enhancing confidence in subsequent docking results.

4.2.11   Case Studies: Receptor Preparation in Drug Discovery

Incorporating case studies showcasing the significance of proper receptor preparation can illustrate the impact on drug discovery. Examples may highlight instances where accurate receptor preparation led to successful identification of binding sites or improved ligand binding affinities.

4.2.12   Machine Learning Applications in Receptor Preparation

The integration of machine learning techniques into receptor preparation is an evolving area. Machine learning models can assist in predicting optimal protonation states, identifying binding sites, or refining protein structures, streamlining the receptor preparation pipeline.

4.2.13   Challenges and Considerations in Receptor Preparation

Despite advancements, challenges exist in receptor preparation, including the accurate handling of membrane proteins, the incorporation of post-translational modifications, and addressing the impact of protein flexibility. Ongoing research aims to enhance methods to address these challenges and improve the reliability of receptor preparation protocols.

4.2.14   Integration with Ligand Preparation

The preparation of receptors is closely linked with the preparation of ligands. Coordinated efforts in optimizing ligand and receptor structures ensure a realistic representation of molecular interactions. Integration of ligand and receptor preparation is vital for the success of molecular docking studies.

4.2.15   Future Directions in Receptor Preparation

The future of receptor preparation involves continuous refinement of methods and increased integration with experimental data. Advances in cryo-electron microscopy, artificial intelligence, and high-throughput screening techniques are expected to shape the landscape of receptor preparation in the coming years.

Chapter 5

Validation and Evaluation

5.1 Validation Metrics

Validation metrics play a pivotal role in assessing the performance and accuracy of molecular docking simulations. Properly chosen metrics provide quantitative measures of how well the predicted ligand binding poses align with experimental data. This section delves into various validation metrics and their application in evaluating the success of molecular docking studies.

5.1.1 Root Mean Square Deviation (RMSD)

The RMSD is a fundamental metric for quantifying the deviation between predicted and experimental ligand poses. Mathematically, it is defined as:

$$\text{RMSD} = \sqrt{\frac{1}{N} \sum_{i=1}^{N} (\mathbf{r}_i - \mathbf{r}_i')^2} \tag{5.1}$$

where $\mathbf{r}_i$ and $\mathbf{r}_i'$ represent the coordinates of atom i in the experimental and predicted poses, respectively. A lower RMSD indicates better alignment.

5.1.2   Ligand Efficiency

Ligand efficiency metrics focus on the relationship between ligand binding affinity and molecular weight. One common form is Ligand Efficiency (LE), given by:

$$LE = \frac{\text{Binding Affinity (kcal/mol)}}{\text{Heavy Atom Count}} \tag{5.2}$$

LE helps prioritize ligands with optimal binding affinities relative to their molecular weight, aiding in lead compound selection.

5.1.3   Area Under the Receiver Operating Characteristic (AUROC)

For studies involving binary classification (active or inactive), AUROC is a widely used metric. It assesses the ability of the docking algorithm to discriminate between true positives and false positives across different thresholds. An ideal model has an AUROC of 1, while random performance yields an AUROC of 0.5.

5.1.4   Enrichment Factor (EF)

Enrichment Factor evaluates the success of a docking protocol in retrieving active compounds early in the ranking. It is defined as:

$$EF = \frac{\text{Fraction of Actives in Top } x}{\text{Fraction of Actives in Entire Dataset}} \tag{5.3}$$

where x is a specified percentage of the dataset. Higher EF values indicate better enrichment.

5.1.5   Consensus Scoring

Consensus scoring combines multiple scoring functions to improve prediction accuracy. Mathematically, the consensus score (CS) is often computed as the weighted sum:

$$CS = \sum_{i=1}^{n} w_i \times \text{Score}_i \tag{5.4}$$

where Score_i is the individual scoring function, and w_i represents its weight.

5.1.6 Specificity and Sensitivity

Specificity (SP) and Sensitivity (SN) are crucial for assessing the ability of a docking protocol to correctly identify true negatives and true positives, respectively. They are defined as:

$$SP = \frac{\text{True Negatives}}{\text{True Negatives} + \text{False Positives}} \tag{5.5}$$

$$SN = \frac{\text{True Positives}}{\text{True Positives} + \text{False Negatives}} \tag{5.6}$$

5.1.7 Kendall's Tau

Kendall's Tau evaluates the correlation between the ranking of ligands by the docking algorithm and their experimental binding affinities. The formula is given by:

$$\tau = \frac{(\text{Number of Concordant Pairs} - \text{Number of Discordant Pairs})}{\frac{1}{2} \times n \times (n-1)} \tag{5.7}$$

where n is the total number of ligands.

5.1.8 Receiver Operating Characteristic (ROC) Curve

The ROC curve is a graphical representation of the true positive rate against the false positive rate at various classification thresholds. The area under the ROC curve (AUROC) is a summary metric, with a higher AUROC indicating better discriminatory power.

5.1.9 Predictive Index (PI)

The Predictive Index evaluates the overall performance of a docking protocol. It combines sensitivity, specificity, and accuracy:

$$PI = \frac{TP + TN}{TP + TN + FP + FN} \tag{5.8}$$

where TP, TN, FP, and FN are true positives, true negatives, false positives, and false negatives, respectively.

5.1.10 Success Rate

The success rate measures the proportion of correctly predicted binding poses within a specified RMSD threshold. It is calculated as:

$$\text{Success Rate} = \frac{\text{Number of Poses with RMSD } < \text{Threshold}}{\text{Total Number of Poses}} \tag{5.9}$$

5.1.11 F-Measure

F-Measure combines precision and recall, providing a single metric for binary classification. It is defined as:

$$F = \frac{2 \times \text{Precision} \times \text{Recall}}{\text{Precision} + \text{Recall}} \tag{5.10}$$

5.1.12 Matthews Correlation Coefficient (MCC)

MCC considers true and false positives and negatives, providing a balanced metric for binary classification:

$$MCC = \frac{TP \times TN - FP \times FN}{\sqrt{(TP + FP)(TP + FN)(TN + FP)(TN + FN)}} \tag{5.11}$$

5.1.13 Distance-Dependent Metrics

Some metrics, like the Fraction of Correctly Predicted Contacts (FCPC) and Fraction of Incorrectly Predicted Contacts (FIPC), assess the accuracy of predicted ligand-receptor contacts at different distances.

5.1.14 Case Studies: Application of Validation Metrics

Incorporating case studies exemplifies the practical application of validation metrics in evaluating the performance of molecular docking protocols. These examples can showcase scenarios where specific metrics elucidate the strengths and limitations of different docking approaches.

5.1.15 Machine Learning for Metric Optimization

Machine learning techniques, including regression models, can be employed to optimize docking metrics. Such models can learn from known ligand-receptor interactions and predict the most relevant metrics for accurate performance assessment.

5.1.16 Limitations and Considerations in Metric Selection

Careful consideration is necessary when selecting validation metrics, as no single metric captures the entire complexity of ligand-receptor interactions. Researchers should choose metrics based on the specific goals of their studies and be aware of the limitations and assumptions inherent to each.

5.1.17 Integration with Experimental Data

Validation metrics are most meaningful when integrated with experimental data. Comparing computational predictions with experimental results provides a holistic understanding of the predictive power of molecular docking studies.

5.1.18   Future Directions in Validation Metrics

Advancements in validation metrics are ongoing, with a focus on developing metrics that address the challenges posed by protein flexibility, solvent effects, and diverse ligand chemistries. The future holds promise for more sophisticated metrics that better align with the complexities of real-world biological systems.

5.2   Benchmark Datasets

Benchmark datasets play a pivotal role in the validation and evaluation of molecular docking algorithms. They provide standardized sets of ligands and receptors with known binding affinities, allowing for a systematic assessment of the predictive capabilities of different docking methods. This section explores the importance of benchmark datasets, their characteristics, and their application in advancing the field of molecular docking.

5.2.1   Purpose of Benchmark Datasets

The primary purpose of benchmark datasets is to facilitate fair and objective comparisons between different docking algorithms. By employing common datasets, researchers can assess the strengths and weaknesses of various methods under consistent conditions, fostering the development of more accurate and reliable docking tools.

5.2.2   Characteristics of Ideal Benchmark Datasets

An ideal benchmark dataset should possess several key characteristics to ensure its effectiveness in evaluating molecular docking algorithms:

- **Diversity:** The dataset should cover a broad range of ligands and receptors, representing various chemical classes and binding site characteristics.

- **Experimental Accuracy:** The binding affinities and poses in the dataset should be experimentally determined with high accuracy, serving as reliable ground truth for validation.

- **Relevance to Drug Discovery:** Benchmark datasets should prioritize ligands with pharmaceutical relevance, ensuring that the assessed algorithms align with real-world drug discovery scenarios.

- **Accessibility:** Open access to benchmark datasets encourages transparency and reproducibility, allowing researchers worldwide to use the same datasets for validation.

5.2.3 Commonly Used Benchmark Datasets

Several benchmark datasets have gained prominence in the field of molecular docking. Examples include:

- **Astex Diverse Set:** A diverse set of protein-ligand complexes with experimentally determined structures and binding affinities.

- **PDBbind:** A comprehensive database containing protein-ligand complexes with known 3D structures and binding affinities, covering a wide range of protein families.

- **DUD-E:** The Directory of Useful Decoys - Enhanced is a benchmark dataset specifically designed for evaluating virtual screening methods, providing decoy structures along with active ligands.

- **CASF-2016:** The Critical Assessment of Structure Prediction (CASP) provides a set of diverse protein-ligand complexes for evaluating ligand binding mode prediction.

5.2.4 Metrics for Benchmark Evaluation

When using benchmark datasets, various metrics can be employed to evaluate the performance of molecular docking algorithms:

- **Success Rate:** The proportion of correctly predicted binding poses within a specified RMSD threshold.

- **Enrichment Factor (EF):** Evaluates the ability of a docking protocol to retrieve active compounds early in the ranking.

- **Receiver Operating Characteristic (ROC) Curve:** Graphical representation of true positive rate against false positive rate at various classification thresholds.

- **Area Under the Curve (AUC):** Quantifies the overall performance of a ROC curve.

5.2.5   Case Studies: Benchmark Dataset Applications

Incorporating case studies that utilize benchmark datasets provides practical insights into the strengths and limitations of different molecular docking algorithms. These examples showcase the real-world applicability of benchmark datasets in informing drug discovery efforts.

5.2.6   Creation of Customized Benchmark Datasets

In addition to using existing benchmark datasets, researchers often create customized datasets tailored to specific research questions or scenarios. This approach allows for the inclusion of unique challenges and considerations relevant to particular drug discovery projects.

5.2.7   Machine Learning in Benchmark Dataset Selection

Machine learning techniques can aid in the selection or creation of benchmark datasets by identifying relevant features, such as ligand-receptor interactions or binding site characteristics. This integration enhances the efficiency of dataset curation.

5.2.8 Limitations of Benchmark Datasets

While benchmark datasets are invaluable tools, they are not without limitations. Challenges include the potential bias introduced by the selection of specific datasets and the difficulty in representing the full complexity of biological systems in a standardized manner.

5.2.9 Integration with Experimental Data

Benchmark datasets achieve their maximum utility when integrated with experimental data. Correlating computational predictions with experimental outcomes enhances the validity and reliability of the assessment, bridging the gap between in silico and in vitro results.

5.2.10 Future Directions in Benchmarking

The continuous evolution of molecular docking algorithms necessitates ongoing refinement of benchmark datasets. Future directions include the incorporation of dynamics information, consideration of membrane proteins, and the development of datasets tailored for specific therapeutic targets.

5.2.11 Community Involvement in Benchmarking

The involvement of the scientific community in benchmarking efforts is crucial for the establishment of standardized practices and the continuous improvement of benchmark datasets. Collaborative initiatives ensure that benchmarking remains a dynamic and adaptive process.

5.2.12 Ensuring Reproducibility in Benchmarking

To maintain the integrity of benchmarking studies, researchers should emphasize transparency and reproducibility. Providing detailed documentation of methodologies, including dataset sources and preprocessing steps, enhances the reliability of benchmarking results.

Chapter 6

Advanced Topics

6.1 Protein-Ligand Interactions

Protein-ligand interactions are pivotal in understanding the molecular basis of various biological processes, including drug binding, enzymatic reactions, and signal transduction. This section delves into the intricacies of protein-ligand interactions, exploring the forces at play, quantifying binding affinities, and examining computational methods to predict and analyze these interactions.

6.1.1 Forces Governing Protein-Ligand Interactions

Protein-ligand interactions are governed by various forces, including:

- **Hydrogen Bonding:** A crucial force where a hydrogen atom of the protein interacts with an electronegative atom of the ligand. Mathematically, the hydrogen bond energy (E_{hb}) can be quantified using the formula:

$$E_{hb} = -k_{hb} \cdot \left(\frac{1}{r_{hb}} - \frac{1}{r_{eq}} \right) \qquad (6.1)$$

where k_{hb} is the force constant, r_{hb} is the current bond length, and r_{eq} is the equilibrium bond length.

- **Van der Waals Forces:** Weak attractive forces arising from transient dipoles in atoms. The van der Waals potential (E_{vdw}) can be modeled using the Lennard-Jones potential:

$$E_{\text{vdw}} = 4\epsilon \left[\left(\frac{\sigma}{r} \right)^{12} - \left(\frac{\sigma}{r} \right)^{6} \right] \qquad (6.2)$$

where ϵ is the depth of the potential well, σ is the finite distance at which the inter-particle potential is zero, and r is the distance between the atoms.

- **Electrostatic Interactions:** Attraction or repulsion between charged atoms in the protein and ligand. The electrostatic energy (E_{elec}) can be calculated using Coulomb's Law:

$$E_{\text{elec}} = \frac{k \cdot q_1 \cdot q_2}{r} \qquad (6.3)$$

where k is Coulomb's constant, q_1 and q_2 are the charges, and r is the distance between the charges.

6.1.2   Quantifying Binding Affinities

The strength of protein-ligand interactions is quantified through various measures:

- **Binding Free Energy (ΔG):** The change in free energy upon binding. It is related to the equilibrium constant (K_{d}) by the equation:

$$\Delta G = -RT \ln K_{\text{d}} \qquad (6.4)$$

where R is the gas constant and T is the temperature.

- **Binding Constant (K_{d}):** The concentration at which half of the protein binding sites are occupied by the ligand. It is inversely proportional to the affinity.

- **Inhibition Constant (K_{i}):** A measure of the ability of a ligand to inhibit a biological process.

6.1.3 Key Binding Motifs

Protein-ligand interactions often exhibit specific binding motifs:

- **Lock-and-Key Model:** The ligand fits precisely into the protein's active site.

- **Induced Fit Model:** The protein undergoes conformational changes upon ligand binding.

- **Hydrogen Bond Networks:** Multiple hydrogen bonds form intricate networks.

- **Salt Bridges:** Electrostatic interactions between charged residues.

- **$\pi - \pi$ Stacking:** Interaction between aromatic rings in the protein and ligand.

6.1.4 Computational Methods for Analysis

Computational methods play a crucial role in the analysis and prediction of protein-ligand interactions:

- **Molecular Docking:** Predicting binding modes and affinities by exploring ligand conformations in the protein's active site.

- **Molecular Dynamics Simulations:** Modeling the dynamic behavior of the protein-ligand complex over time.

- **Quantum Mechanical Calculations:** Understanding electronic structure and energetics.

- **Free Energy Calculations:** Estimating binding free energy using methods like MM/PBSA.

- **Machine Learning Approaches:** Predicting binding affinities based on training datasets.

6.1.5   Example: Hydrogen Bonding in Drug Design

Consider a drug designed to target an enzyme's active site through hydrogen bonding. The drug's functional group may contain a hydrogen bond donor (e.g., NH group) that forms a strong hydrogen bond with a complementary acceptor group in the enzyme (e.g., C=O group).

6.1.6   Challenges in Predicting Interactions

Despite advancements, predicting protein-ligand interactions poses challenges:

- **Conformational Flexibility:** Both proteins and ligands can exhibit dynamic behavior.

- **Water Molecules:** Inclusion of water molecules in the binding site.

- **Entropic Contributions:** Accounting for changes in entropy during binding.

6.1.7   Future Directions in Protein-Ligand Interactions

The future of studying protein-ligand interactions involves:

- **Integration of Experimental and Computational Approaches:** Combining computational predictions with experimental validations.

- **Advancements in Quantum Computing:** Handling complex calculations more efficiently.

- **Personalized Medicine:** Tailoring drug designs based on individual patient profiles.

6.2   Flexible Docking

Flexible docking is a sophisticated approach in molecular docking that considers the dynamic nature of both ligands and receptors during the binding process. In traditional docking studies, rigidity is assumed for both ligands and receptors, neglecting the conformational changes that often occur upon binding. This section explores the significance of flexible docking, the challenges it addresses, and the computational methods employed to model and analyze these dynamic interactions.

6.2.1 Importance of Flexibility in Molecular Docking

In many biological processes, both ligands and receptors undergo conformational changes to achieve optimal binding. Flexible docking recognizes the importance of accounting for this flexibility to improve the accuracy of predicting binding modes and affinities. By allowing for changes in the shape and orientation of the ligand and receptor, flexible docking provides a more realistic representation of the binding process.

6.2.2 Mathematical Modeling of Flexible Docking

Flexible docking involves the incorporation of conformational flexibility into the docking algorithms. This can be achieved through the use of mathematical models that describe the potential energy surface of the ligand-receptor system. One common approach is to employ molecular dynamics simulations to explore the conformational space and identify energetically favorable binding modes.

6.2.3 Scoring Functions in Flexible Docking

Scoring functions play a crucial role in flexible docking by evaluating the fitness of different ligand conformations within the receptor's binding site. These scoring functions consider factors such as van der Waals interactions, electrostatic forces, hydrogen bonding, and solvation effects. A popular example is the Generalized Born Surface Area (GBSA) method, which estimates binding free energies by considering solvation effects.

6.2.4 Handling Protein Flexibility

Protein flexibility is a key aspect of flexible docking. Techniques like induced fit docking explicitly account for protein conformational changes upon ligand binding. This involves multiple steps, starting with an initial rigid docking, followed by protein side-chain and backbone adjustments to accommodate the ligand.

6.2.5 Example: Induced Fit Docking

Consider a ligand binding to a receptor with a flexible binding site. In induced fit docking, the initial docking is performed with a rigid receptor, generating a set of potential ligand poses. The best poses are then selected, and the receptor is subjected to further optimization to accommodate the ligand's conformation. This iterative process refines the binding mode, considering both ligand and receptor flexibility.

6.2.6 Incorporating Ligand Flexibility

Ligand flexibility is equally crucial in flexible docking studies. The ligand may adopt different conformations to maximize interactions with the receptor. Flexible ligand docking methods explore the ligand's conformational space to identify energetically favorable binding modes.

6.2.7 Mathematical Formulation of Ligand Flexibility

Mathematically, ligand flexibility can be represented using methods such as molecular dynamics simulations or Monte Carlo sampling. The ligand's potential energy is calculated as a function of its internal coordinates, allowing for the exploration of different conformations. The Metropolis criterion is often employed to accept or reject proposed conformational changes based on energy differences.

6.2.8 Hybrid Approaches in Flexible Docking

Hybrid approaches combine aspects of rigid and flexible docking to strike a balance between computational efficiency and accuracy. These methods may involve initial rigid docking followed by refinement steps that consider flexibility in either the ligand, receptor, or both.

6.2.9 Challenges in Flexible Docking

Despite its advantages, flexible docking poses several challenges. The increased computational complexity and the need for accurate force fields for both ligands and receptors are critical considerations. Efficient sampling of the conformational space without missing relevant binding modes is another challenge.

6.2.10 Advanced Sampling Techniques

To overcome challenges in sampling, advanced techniques such as enhanced sampling methods or machine learning-based approaches can be employed. These techniques aim to explore the conformational space more efficiently, improving the accuracy of flexible docking predictions.

6.2.11 Validation and Benchmarking of Flexible Docking

Assessing the performance of flexible docking methods requires validation against experimental data and benchmarking against established datasets. Metrics such as RMSD (Root Mean Square Deviation) of predicted binding poses, enrichment factors, and binding free energy correlations are commonly used for benchmarking.

6.2.12 Integration with Experimental Data

Validating flexible docking predictions with experimental data is crucial for establishing the reliability of the results. Experimental techniques such as X-ray crystallography, NMR spectroscopy, or biochemical assays can confirm the predicted binding modes and strengths.

6.2.13   Applications in Drug Discovery

Flexible docking has found widespread applications in drug discovery. By capturing the dynamic nature of ligand-receptor interactions, it aids in the design of more potent and selective drug candidates. Understanding how ligands adapt to different receptor conformations is vital for optimizing binding affinities.

6.2.14   Future Directions in Flexible Docking

The future of flexible docking involves further refinement of sampling techniques, incorporation of more accurate force fields, and exploration of allosteric binding sites. Collaborative efforts between computational biologists, chemists, and experimentalists will drive advancements in this field.

Chapter 7

Applications

7.1 Drug Discovery

Drug discovery is a multifaceted process that aims to identify and develop new therapeutic compounds to treat diseases. Molecular docking plays a crucial role in this process by predicting the binding modes and affinities of potential drug candidates with target proteins. This section explores the application of molecular docking in drug discovery, highlighting its significance, challenges, and examples of successful endeavors.

7.1.1 Significance of Molecular Docking in Drug Discovery

Molecular docking serves as a powerful tool in drug discovery for several reasons:

- **Rational Drug Design:** Docking allows for the rational design of new drugs by predicting how potential compounds interact with target proteins at the molecular level.

- **Virtual Screening:** High-throughput virtual screening of compound libraries accelerates the identification of lead compounds with favorable binding properties.

- **Understanding Binding Mechanisms:** Docking provides insights into the specific interactions between drugs and their target proteins, aiding in the optimization of binding affinities.

- **Reducing Experimental Costs:** Prioritizing potential drug candidates through computational methods reduces the number of compounds that need to be synthesized and tested experimentally.

7.1.2   Mathematical Formulation of Binding Affinity

The binding affinity (K_d) of a drug to its target protein is a critical parameter in drug discovery. It can be calculated using the following equation:

$$K_d = \frac{k_{off}}{k_{on}} \qquad (7.1)$$

where k_{off} is the dissociation rate constant, and k_{on} is the association rate constant. A lower K_d indicates a stronger binding affinity.

7.1.3   Examples of Successful Drug Discovery Through Docking

Several drugs have been discovered or optimized using molecular docking:

- **Tamiflu (Oseltamivir):** Used to treat influenza, Tamiflu was optimized using docking studies to improve its binding to the neuraminidase enzyme.

- **Raloxifene:** Originally developed for osteoporosis, molecular docking played a role in repositioning Raloxifene as a potential treatment for breast cancer.

- **Zafirlukast:** A leukotriene receptor antagonist for asthma treatment, optimized through docking studies to enhance its binding to the target receptor.

7.1.4 Challenges in Molecular Docking for Drug Discovery

Despite its success, molecular docking faces challenges in the context of drug discovery:

- **Scoring Function Accuracy:** The accuracy of scoring functions in predicting binding affinities is a persistent challenge, requiring ongoing refinement.

- **Treatment of Flexibility:** Accounting for protein and ligand flexibility is crucial for accurate predictions but adds computational complexity.

- **Water Molecules and Solvation:** The role of water molecules in binding sites and solvation effects presents challenges in docking accuracy.

7.1.5 Integration with Experimental Validation

While molecular docking is a valuable tool, experimental validation is essential for confirming predicted interactions. Techniques such as X-ray crystallography, NMR spectroscopy, and biochemical assays provide crucial data to validate computational predictions.

7.1.6 Combining Docking with ADMET Prediction

To enhance the drug discovery process, molecular docking can be integrated with ADMET (Absorption, Distribution, Metabolism, Excretion, and Toxicity) prediction models. This integrated approach helps assess the drug-likeness and safety profile of potential candidates.

7.1.7 Application of Machine Learning in Drug Discovery

Machine learning techniques are increasingly applied in drug discovery to predict binding affinities, optimize lead compounds, and analyze large datasets. These methods complement molecular docking, offering a data-driven approach to drug design.

7.1.8   Case Study: Designing a Kinase Inhibitor

Consider the design of a kinase inhibitor using molecular docking. By exploring the conformational space of potential inhibitors and predicting their binding modes, researchers can optimize lead compounds for maximum binding affinity and specificity.

7.1.9   Structure-Based Drug Design

Molecular docking forms the basis of structure-based drug design, where the three-dimensional structure of the target protein guides the design of new drug molecules. This approach has been successful in the development of various therapeutic agents.

7.1.10   Fragment-Based Drug Design

Fragment-based drug design involves docking smaller molecular fragments into the target site, gradually building a complete drug molecule. This approach is efficient for exploring chemical space and identifying high-affinity binding fragments.

7.1.11   Future Directions in Drug Discovery with Docking

The future of drug discovery with molecular docking involves:

- **Improved Scoring Functions:** Advancements in scoring functions for more accurate prediction of binding affinities.

- **Incorporation of Machine Learning:** Further integration of machine learning models for enhanced predictions.

- **Personalized Medicine:** Tailoring drug treatments based on individual genetic profiles for increased efficacy.

7.2 Virtual Screening

Virtual screening is a computational approach employed in drug discovery to sift through large compound libraries and identify potential drug candidates that may exhibit favorable interactions with target proteins. Molecular docking plays a central role in virtual screening, predicting the binding affinity of ligands to target proteins. This section delves into the principles of virtual screening, its methodologies, and applications in accelerating the drug discovery process.

7.2.1 Principles of Virtual Screening

The fundamental principles of virtual screening include:

- **Ligand-Based Screening:** Utilizing known ligands to identify structurally similar compounds in large databases.

- **Structure-Based Screening:** Employing the three-dimensional structure of the target protein to predict the binding affinity of potential ligands.

- **Pharmacophore-Based Screening:** Identifying common features essential for ligand binding and searching for compounds that match these features.

7.2.2 Mathematical Formulation of Screening Metrics

Virtual screening involves various metrics to assess the likelihood of ligand binding. One common metric is the docking score (S_{dock}), calculated by the scoring function. The higher the docking score, the more favorable the interaction. It is expressed as:

$$S_{\text{dock}} = w_1 E_{\text{vdw}} + w_2 E_{\text{elec}} + w_3 E_{\text{hb}} + \ldots \tag{7.2}$$

where E_{vdw}, E_{elec}, E_{hb}, and others represent van der Waals, electrostatic, and hydrogen bonding energies, respectively. The weights w_1, w_2, w_3, etc., are parameters adjusted to optimize the scoring function.

7.2.3   Methodologies in Virtual Screening

Virtual screening methodologies can be broadly categorized into:

- **High-Throughput Docking:** Screening a large number of compounds against a target protein using automated docking algorithms.

- **Pharmacophore Modeling:** Defining the essential features a ligand must possess for binding, and using these features to search databases.

- **Quantitative Structure-Activity Relationship (QSAR):** Establishing a mathematical relationship between chemical structure and biological activity, facilitating prediction.

7.2.4   Applications of Virtual Screening

Virtual screening finds applications in various stages of drug discovery:

- **Lead Identification:** Identifying potential lead compounds with high binding affinities.

- **Lead Optimization:** Refining lead compounds to enhance their binding properties.

- **Repositioning:** Exploring existing drugs for new therapeutic applications.

- **Polypharmacology:** Investigating compounds that can interact with multiple targets.

7.2.5   Case Study: Virtual Screening for Antiviral Agents

Consider the virtual screening for antiviral agents targeting a specific viral protein. By screening a large compound library, potential inhibitors can be identified, and their binding affinities predicted through molecular docking. This accelerates the discovery of new antiviral drugs.

7.2.6 Integration with Experimental Validation

While virtual screening expedites the identification of potential drug candidates, experimental validation is indispensable. Techniques such as biochemical assays, X-ray crystallography, and in vivo studies are crucial for confirming the predicted interactions.

7.2.7 Challenges in Virtual Screening

Despite its utility, virtual screening faces challenges:

- **Scoring Function Accuracy:** Similar to docking, the accuracy of scoring functions in predicting binding affinities is a continual challenge.

- **Treatment of Protein Flexibility:** Accounting for protein flexibility during screening to improve accuracy.

- **Library Size and Diversity:** The quality and diversity of compound libraries impact the success of virtual screening.

7.2.8 Emerging Technologies in Virtual Screening

Advancements in technology continue to enhance virtual screening:

- **Machine Learning Integration:** Utilizing machine learning models for improved prediction accuracy.

- **Quantum Computing:** Exploring the potential of quantum computing for more efficient virtual screening.

- **Cloud-Based Platforms:** Enabling researchers to access high-performance computing resources for large-scale virtual screening.

7.2.9 Future Directions in Virtual Screening

The future of virtual screening involves:

- **Advancements in Scoring Functions:** Development of more accurate scoring functions through advanced mathematical models.

- **Personalized Virtual Screening:** Tailoring virtual screening approaches based on individual patient profiles.

- **Integration with Systems Biology:** Considering network-level interactions in virtual screening.

7.3   Case Studies

This section presents several case studies highlighting the practical applications of molecular docking in various domains, showcasing its versatility in drug discovery, protein-ligand interactions, and structure-based design.

7.3.1   Drug Repurposing: Rediscovering Famotidine

Drug repurposing, also known as drug repositioning, has gained significant attention as a strategy to identify new therapeutic uses for existing drugs. Famotidine, originally developed as a histamine-2 receptor antagonist for treating peptic ulcers and gastroesophageal reflux disease (GERD), has emerged as a promising candidate for repurposing due to its potential antiviral properties.

Antiviral Potential of Famotidine

Recent studies have suggested that famotidine may exhibit antiviral effects, particularly against RNA viruses. While the exact mechanisms are still under investigation, it is believed that famotidine might interfere with viral replication processes. This has sparked interest in exploring its efficacy against various viral infections, including those caused by RNA viruses such as influenza and coronaviruses.

Chemical Formulas

Representing the repurposing of famotidine in chemical formulas:

$$\text{Original Use:} \quad \text{Famotidine} \longrightarrow \text{Peptic Ulcers} + \text{GERD} \qquad (7.3)$$

$$\text{Repurposing:} \quad \text{Famotidine} \longrightarrow \text{Antiviral Applications} \qquad (7.4)$$

These chemical formulas illustrate the dual role of famotidine in its original therapeutic applications and its potential repurposing for antiviral purposes.

Clinical Trials and Investigations

Ongoing clinical trials and investigations are evaluating famotidine's efficacy in the context of viral infections. Researchers are exploring its impact on viral load, disease progression, and overall patient outcomes. Results from these studies will contribute to a better understanding of famotidine's repurposing potential.

Future Implications

The repurposing of famotidine represents a paradigm shift in drug discovery, showcasing the versatility of existing medications. If successful, it could offer a cost-effective and expedited approach to finding treatments for viral infections. However, further research and clinical validation are necessary to establish famotidine's efficacy and safety in antiviral applications.

The exploration of famotidine's repurposing exemplifies the dynamic nature of drug discovery and the potential for existing drugs to address unmet medical needs in unforeseen ways.

7.3.2   Designing Selective Kinase Inhibitors

Kinases play a crucial role in cellular signaling pathways, and their dysregulation is often associated with various diseases, including cancer. Designing selective kinase inhibitors is a key focus in drug discovery, aiming to develop therapeutic agents that specifically target aberrant kinase activity while minimizing off-target effects.

Importance of Selectivity

Achieving selectivity in kinase inhibition is essential to avoid unintended interactions with other kinases and off-target proteins. Non-selective inhibitors may
lead to adverse effects and limit the therapeutic potential of a drug. There-
fore, rational drug design strategies focus on enhancing the selectivity profile of
kinase inhibitors.

Mathematical Formulas

Quantifying selectivity using mathematical formulas:

$$\text{Selectivity Index (SI)} = \frac{\text{Activity against Target Kinase}}{\text{Activity against Off-Target Kinase}} \qquad (7.5)$$

Here, the Selectivity Index is a ratio that provides a quantitative measure
of a kinase inhibitor's preference for the target kinase over an off-target kinase.
A higher SI indicates greater selectivity.

Structural Considerations

The three-dimensional structure of kinases plays a crucial role in designing selective inhibitors. Molecular docking studies and structure-activity relationship
(SAR) analyses help identify key binding interactions and structural features
that contribute to selectivity.

Case Studies

Case studies of successfully designed selective kinase inhibitors, such as imatinib
for BCR-ABL in chronic myeloid leukemia (CML) and vemurafenib for BRAF in
melanoma, showcase the impact of selective kinase targeting in clinical settings.

Emerging Technologies

Advancements in computational approaches, including machine learning and
artificial intelligence, are aiding in the prediction of kinase selectivity profiles.

These technologies accelerate the drug discovery process by predicting potential off-target interactions and optimizing selectivity during the early stages of inhibitor design.

Challenges and Future Directions

Despite significant progress, achieving absolute selectivity remains challenging. Kinase inhibitor design faces obstacles such as the dynamic nature of kinase conformations and the potential for resistance mutations. Future directions involve exploring innovative strategies, including allosteric inhibition and covalent binding, to enhance selectivity and overcome existing challenges.

Designing selective kinase inhibitors represents a dynamic and evolving field within drug discovery, with the potential to significantly impact the treatment of various diseases associated with kinase dysregulation.

7.3.3 Optimizing Antibody-Drug Conjugates

Antibody-Drug Conjugates (ADCs) have emerged as a promising class of therapeutics that combine the specificity of monoclonal antibodies with the potency of cytotoxic drugs. Optimizing ADCs involves addressing various aspects, including antibody selection, linker design, and payload selection, to enhance their efficacy and safety profiles.

Antibody Selection

The choice of antibody is a critical factor in ADC optimization. Selection criteria include high binding affinity to the target antigen, internalization efficiency upon binding, and minimal immunogenicity. Advances in antibody engineering, such as the development of humanized or fully human antibodies, contribute to improved therapeutic outcomes.

Linker Design

The linker that connects the antibody and the drug payload plays a pivotal role in ADC stability and drug release kinetics. Balancing stability during circulation and efficient payload release upon internalization is crucial. Various linker types, including cleavable and non-cleavable linkers, are employed based on the desired release mechanism.

Payload Selection

Optimal payload selection involves choosing a cytotoxic drug with potent anti-tumor activity. Factors such as drug potency, stability, and the ability to induce cell death upon release contribute to payload selection. Additionally, the hydrophobicity of the payload influences ADC solubility and pharmacokinetics.

Mathematical Formulas

Quantitative metrics for ADC optimization:

$$\text{Therapeutic Index (TI)} = \frac{\text{Maximum Tolerated Dose (MTD) in vivo}}{\text{Minimum Effective Dose (MED) in vivo}} \tag{7.6}$$

The Therapeutic Index is a crucial parameter that reflects the balance between the maximum tolerated dose and the minimum effective dose, providing a quantitative measure of ADC efficacy and safety.

Structural Considerations

Understanding the three-dimensional structure of ADC components is essential for rational design. Molecular modeling and computational simulations aid in predicting potential interactions, optimizing linker stability, and evaluating the conformational flexibility of ADCs.

Case Studies

Successful optimization of ADCs is exemplified by the clinical approval of trastuzumab emtansine (T-DM1) for HER2-positive breast cancer. T-DM1 incorporates a stable linker and the microtubule inhibitor DM1, showcasing the impact of rational design on therapeutic success.

Future Directions

Ongoing research focuses on advancing ADC technology, including the development of next-generation linkers, novel payload classes, and strategies to overcome resistance mechanisms. Personalized approaches, tailoring ADCs to specific patient populations or tumor subtypes, represent an exciting avenue for future exploration.

Optimizing Antibody-Drug Conjugates continues to be a dynamic field, with innovations aimed at improving therapeutic outcomes and expanding the application of ADCs in cancer and beyond.

7.3.4 Understanding Enzyme-Substrate Interactions: Trypsin-Catalyzed Reactions

Enzyme-substrate interactions are fundamental to biochemical processes, and studying specific examples, such as trypsin-catalyzed reactions, provides insights into the intricacies of enzymatic mechanisms.

Introduction to Trypsin

Trypsin is a serine protease that plays a crucial role in the digestion of proteins. It cleaves peptide bonds on the carboxyl side of lysine and arginine residues, contributing to the breakdown of polypeptides into smaller peptides.

Enzyme-Substrate Recognition

The specificity of trypsin for certain peptide bonds is central to its catalytic function. Trypsin recognizes and binds to its substrate through interactions between amino acid residues in the enzyme's active site and specific side chains of the substrate.

Catalytic Mechanism

The catalytic mechanism of trypsin involves a nucleophilic attack on the carbonyl carbon of the peptide bond by the serine residue in the enzyme's active site. This results in the formation of a covalent acyl-enzyme intermediate, which is subsequently hydrolyzed to release the cleaved peptide.

Mathematical Representation

The rate equation for the trypsin-catalyzed reaction can be represented as follows:

$$\text{Rate} = k \cdot [\text{Substrate}] \cdot [\text{Enzyme}] \tag{7.7}$$

Here, [Substrate] and [Enzyme] denote the concentrations of substrate and enzyme, respectively, and k is the rate constant.

Experimental Techniques

Studying trypsin-catalyzed reactions often involves techniques such as enzyme kinetics assays, X-ray crystallography, and nuclear magnetic resonance (NMR) spectroscopy. These methods provide data on reaction rates, enzyme-substrate complex structures, and dynamic changes during catalysis.

Application in Biotechnology

Understanding trypsin-catalyzed reactions has practical applications in biotechnology, particularly in protein purification and cell culture. Trypsin is commonly used to detach cells from culture vessels during cell passaging.

Case Studies

Case studies exploring specific substrates and inhibitors of trypsin shed light on the versatility of this enzyme. Insights gained from these studies contribute to the development of targeted drugs and therapeutic interventions.

Future Directions

Ongoing research aims to further unravel the molecular details of trypsin-catalyzed reactions and explore potential applications in drug discovery. Advances in computational methods and structural biology continue to enhance our understanding of enzyme-substrate interactions.

Studying trypsin-catalyzed reactions provides a model system for comprehending broader principles of enzymology and serves as a foundation for advancing our understanding of enzyme-substrate interactions in diverse biological contexts.

7.3.5 Targeting G-Protein Coupled Receptors (GPCRs) in Drug Discovery

G-Protein Coupled Receptors (GPCRs) represent a crucial class of cell surface receptors involved in signal transduction. Targeting GPCRs in drug discovery has been a significant area of research, given their central role in mediating responses to various physiological signals.

Introduction to GPCRs

GPCRs are membrane proteins that transmit signals from the extracellular environment to the interior of cells. They play a key role in regulating various physiological processes, making them attractive targets for drug development.

Diversity of GPCR Ligands

The ligands that interact with GPCRs are diverse, ranging from small molecules to peptides and large proteins. Understanding the binding specificity of ligands to GPCRs is crucial for designing drugs that modulate their activity.

GPCR Signaling Pathways

Upon ligand binding, GPCRs can activate intracellular signaling pathways through the modulation of G-proteins. These pathways regulate diverse cellular functions, including neurotransmission, immune response, and hormonal regulation.

Mathematical Modeling of GPCR Activation

Mathematical models, such as the ternary complex model, are employed to describe the binding and activation of GPCRs. These models incorporate parameters related to ligand-receptor interactions and downstream signaling events.

Drug Discovery Strategies

Drug discovery efforts targeting GPCRs involve high-throughput screening, structure-based drug design, and virtual screening. Advances in computational techniques and structural biology have facilitated the identification of novel GPCR modulators.

Case Studies in GPCR Drug Discovery

Exploring case studies of successful GPCR-targeted drugs, such as beta-blockers and antipsychotics, provides insights into the challenges and opportunities in this field. Understanding the structural basis of ligand-GPCR interactions is crucial for rational drug design.

Biased Signaling and Allosteric Modulation

Recent research has focused on biased signaling, where ligands selectively activate specific signaling pathways. Additionally, allosteric modulation of GPCRs offers alternative strategies for drug development, providing more options for therapeutic intervention.

Emerging Technologies

Advancements in cryo-electron microscopy (cryo-EM) and other structural biology techniques have accelerated our understanding of GPCR structures, enabling more precise drug design. Integrating these technologies with computational approaches enhances the drug discovery process.

Future Perspectives

The continued exploration of GPCRs in drug discovery holds promise for developing innovative therapies across various disease areas. Targeting specific GPCRs with tailored ligands offers a personalized approach to pharmacotherapy.

Understanding the intricacies of GPCR signaling and employing cutting-edge technologies will contribute to the development of safer and more effective drugs. GPCRs remain a dynamic and evolving field in drug discovery, driving advancements in precision medicine.

7.3.6 Virtual Screening for Anti-Malarial Compounds

Malaria, a life-threatening disease caused by Plasmodium parasites, remains a global health challenge. Virtual screening, a computational approach, has emerged as a valuable tool in the search for novel anti-malarial compounds.

Computational Methods in Virtual Screening

Virtual screening involves the use of computational methods to predict the binding affinity of small molecules to a target, typically a biomolecular structure. Key computational techniques include molecular docking, molecular dynamics simulations, and pharmacophore modeling.

Molecular Docking for Anti-Malarial Drug Discovery

Molecular docking is a widely used virtual screening technique that predicts the preferred orientation and binding affinity of a ligand to a target protein. The scoring function in molecular docking helps identify potential anti-malarial compounds based on their interaction with specific biomolecular targets.

$$\text{Scoring Function: Score} = \text{Weight}_{\text{vdW}} \cdot \text{vdW} + \text{Weight}_{\text{elec}} \qquad (7.8)$$
$$\cdot \text{ elec} + \text{Weight}_{\text{desolv}} \cdot \text{desolv}$$

Here, vdW represents van der Waals interactions, elec represents electrostatic interactions, and desolv represents desolvation energy.

Pharmacophore Modeling for Anti-Malarial Compounds

Pharmacophore modeling involves identifying essential molecular features that contribute to the biological activity of a ligand. In the context of anti-malarial drug discovery, pharmacophore models help filter and prioritize compounds based on their potential to interact with specific parasite targets.

$$\text{Pharmacophore Features: Hydrogen Bond Donor,} \qquad (7.9)$$
$$\text{Hydrogen Bond Acceptor, Aromatic Ring,} \ldots$$

Molecular Dynamics Simulations

Molecular dynamics simulations provide insights into the dynamic behavior of biomolecular systems over time. In the context of anti-malarial drug discovery, these simulations help understand the stability and flexibility of ligand-protein complexes, aiding in the selection of promising compounds.

Case Studies in Virtual Screening for Malaria

Exploring case studies of virtual screening campaigns for anti-malarial compounds sheds light on successful strategies and challenges faced by researchers. Examples include the identification of compounds targeting key enzymes involved in the parasite life cycle.

Integration of Machine Learning

Machine learning algorithms, such as random forests or support vector machines, are increasingly integrated into virtual screening pipelines to enhance predictive models. These algorithms learn from known anti-malarial compounds and predict the likelihood of new compounds being effective.

$$\text{Machine Learning Model: Prediction} = \text{RF}(\text{Physicochemical Properties, Structural Features}, \ldots) \tag{7.10}$$

Chemical Formulas

Chemical formulas for potential anti-malarial compounds may include structures targeting specific parasite biomolecules. For example:

$$C_{17}H_{21}ClN_4O \quad \text{(Chemical Structure of a Virtual Hit)} \tag{7.11}$$

Challenges and Future Directions

Despite advancements, challenges in virtual screening for anti-malarial compounds persist. Overcoming issues related to target flexibility, ligand parameterization, and model validation remains crucial. Future directions include the incorporation of artificial intelligence and big data analytics for more robust predictions.

Virtual screening continues to play a pivotal role in accelerating the discovery of novel anti-malarial compounds, offering a computational approach to complement traditional experimental methods.

7.3.7 Rational Design of Antiviral Protease Inhibitors

The rational design of antiviral protease inhibitors represents a crucial strategy in combating viral infections. Proteases play a key role in the viral life cycle, making them attractive targets for drug development.

Targeting Viral Proteases

Viral proteases are enzymes essential for the cleavage of viral polyproteins into functional components. Inhibiting these proteases disrupts viral replication and maturation, making them promising targets for antiviral drug design.

Mathematical Model for Inhibition Kinetics

The kinetics of protease inhibition can be mathematically described using models such as the Michaelis-Menten equation. For a competitive protease inhibitor:

$$V = \frac{V_{\max} \cdot [S]}{K_m \cdot (1 + \frac{[I]}{K_i}) + [S]} \tag{7.12}$$

Here, V is the reaction velocity, $[S]$ is the substrate concentration, $[I]$ is the inhibitor concentration, $V_{\max}$ is the maximum velocity, K_m is the Michaelis constant, and K_i is the inhibition constant.

Structural Basis of Inhibition

Understanding the three-dimensional structure of viral proteases is critical for rational drug design. X-ray crystallography and cryo-electron microscopy provide insights into the binding pocket of proteases and interactions with inhibitors.

Chemical Formulas

Chemical formulas for antiviral protease inhibitors may include structures targeting specific amino acid residues in the protease active site. For example:

$$C_{28}H_{34}N_6O_5S \quad \text{(Chemical Structure of a Protease Inhibitor)} \tag{7.13}$$

Optimizing Pharmacokinetics

In addition to inhibitory potency, optimizing pharmacokinetic properties such as bioavailability and metabolic stability is crucial for the development of effective antiviral protease inhibitors.

Case Studies in Antiviral Drug Design

Examining case studies of successful antiviral protease inhibitors, such as those developed for HIV or HCV, provides valuable insights into rational drug design strategies. This includes the evolution of compounds from initial hits to clinically approved drugs.

Emerging Trends and Future Directions

Advancements in computational methods, including molecular dynamics simulations and machine learning, are shaping the future of antiviral drug design. Predicting the impact of mutations and understanding resistance mechanisms are key challenges for the next generation of protease inhibitors.

The rational design of antiviral protease inhibitors continues to be a dynamic field, driving innovation in drug development and contributing to the arsenal against viral infections.

7.3.8 Optimizing Peptide Ligands for Receptor Binding

The optimization of peptide ligands for receptor binding is a crucial aspect of drug design, especially in the development of peptide-based therapeutics. Achieving high affinity and specificity is essential for the success of peptide drugs.

Mathematical Model for Peptide-Receptor Binding

The binding kinetics between a peptide ligand (P) and its receptor (R) can be described using a mathematical model. The association and dissociation rates (k_{on} and k_{off}) contribute to the overall binding affinity (K_d):

$$K_d = \frac{k_{\mathrm{off}}}{k_{\mathrm{on}}} \tag{7.14}$$

Understanding and manipulating these rates are crucial for optimizing the binding strength of peptide ligands.

Structural Considerations

The three-dimensional structure of a peptide greatly influences its binding to a receptor. Techniques such as nuclear magnetic resonance (NMR) and X-ray crystallography provide insights into the conformation of peptides in complex with their receptors.

Chemical Formulas

Chemical formulas play a vital role in representing the structure of optimized peptide ligands. For example, the chemical structure of an optimized peptide ligand targeting a specific receptor can be represented as:

$$H_2N-CH(CO-NH_2)-Phe-Leu-Arg-CONH_2 \tag{7.15}$$

In Silico Design Strategies

Computational methods, including molecular docking and dynamics simulations, facilitate the in silico optimization of peptide ligands. These strategies help predict the binding affinity and guide modifications for enhanced receptor interactions.

Case Studies in Peptide Drug Development

Exploring case studies of successful peptide drugs, such as those targeting G protein-coupled receptors (GPCRs) or enzymes, provides valuable insights into the optimization process. This includes modifications made to improve pharmacokinetics and reduce off-target effects.

Biophysical Techniques for Binding Analysis

Biophysical techniques, including surface plasmon resonance (SPR) and isothermal titration calorimetry (ITC), enable accurate measurement of binding kinetics and thermodynamics. These techniques contribute to the quantitative assessment of optimized peptide-receptor interactions.

Future Directions and Challenges

The optimization of peptide ligands continues to evolve with advancements in computational and experimental techniques. Overcoming challenges such as stability, immunogenicity, and delivery methods is crucial for the successful translation of optimized peptides into therapeutic applications.

The ongoing efforts to optimize peptide ligands for receptor binding underscore the importance of interdisciplinary approaches in drug design and development.

7.3.9 Structure-Based Design of Anti-Inflammatory Agents

The structure-based design of anti-inflammatory agents involves leveraging knowledge of the three-dimensional structures of target proteins to create molecules with enhanced therapeutic effects. This approach is crucial for developing potent and selective drugs to combat inflammation.

Molecular Docking for Target Identification

Molecular docking plays a pivotal role in the structure-based design process. It involves predicting the preferred orientation of a ligand molecule concerning a protein receptor. The binding affinity (E_{binding}) can be estimated using the following equation:

$$E_{\text{binding}} = E_{\text{ligand}} + E_{\text{receptor}} + E_{\text{interactions}} \tag{7.16}$$

Here, E_{ligand} is the energy of the ligand, E_{receptor} is the energy of the receptor, and $E_{\text{interactions}}$ is the energy of ligand-receptor interactions.

Chemical Formulas

The chemical structures of designed anti-inflammatory agents are represented using chemical formulas. For instance, a nonsteroidal anti-inflammatory drug (NSAID) such as aspirin can be represented as:

$$C_9H_8O_4 \tag{7.17}$$

Optimizing Drug-Receptor Interactions

Optimizing the interactions between the drug and its target receptor is essential. This can involve modifying the chemical structure to enhance hydrogen bonding, hydrophobic interactions, or electrostatic forces. Quantitative structure-activity relationship (QSAR) models aid in predicting how structural changes affect biological activity.

In Silico ADMET Prediction

In silico prediction of absorption, distribution, metabolism, excretion, and toxicity (ADMET) properties is crucial for filtering potential drug candidates. Computational tools estimate parameters such as lipophilicity, solubility, and toxicity, aiding in the selection of compounds with favorable pharmacokinetic profiles.

Case Studies in Anti-Inflammatory Drug Design

Exploring case studies of successful anti-inflammatory drugs, such as selective cyclooxygenase-2 (COX-2) inhibitors or biologics targeting inflammatory cytokines, provides valuable insights into effective design strategies. Understanding the structural basis for their activity guides the development of novel agents.

Biophysical Techniques for Validation

Biophysical techniques, including X-ray crystallography and nuclear magnetic resonance (NMR), validate the binding modes of designed agents. Experimental validation ensures that the predicted binding interactions align with real-world observations.

Future Directions and Challenges

The field of structure-based design for anti-inflammatory agents faces challenges, including predicting off-target effects and optimizing selectivity. Future directions may involve incorporating machine learning approaches to improve the accuracy of predictions and streamline the drug discovery process.

The continuous exploration of structure-based design principles contributes to the development of next-generation anti-inflammatory agents with improved efficacy and reduced side effects.

7.3.10 Exploring Allosteric Modulation in Enzymes

Allosteric modulation in enzymes represents a fascinating area of study, offering unique insights into the regulation of enzymatic activity through binding at sites distinct from the active site. This subsection explores the principles and applications of allosteric modulation in the context of enzymes.

Definition and Mechanism

Allosteric modulation involves the binding of a ligand at a site on an enzyme other than the active site, leading to a conformational change that alters the enzyme's activity. The allosteric site may be spatially distinct from the active site, impacting the enzyme's catalytic properties.

Mathematical Representation

The effect of allosteric modulation can be mathematically represented using the MWC (Monod-Wyman-Changeux) model. For an enzyme with multiple subunits, the model is given by:

$$Y = \frac{Y_0(1 + \epsilon_A[S] + \epsilon_I[I])}{1 + \alpha_A[S] + \beta_A[S]^2 + \gamma_I[I]} \tag{7.18}$$

Here, Y represents the enzyme activity, ϵ_A, ϵ_I, α, β, and γ are constants, and $[S]$ and $[I]$ are the concentrations of substrate and allosteric modulator, respectively.

Allosteric Enzyme Inhibition

Allosteric modulation can lead to enzyme inhibition. The Hill equation can describe the relationship between the fractional saturation (Y/Y_0) and the concentration of the allosteric modulator:

$$Y/Y_0 = \frac{[I]^n}{(K_i^n + [I]^n)} \qquad (7.19)$$

Here, n is the Hill coefficient, and K_i is the allosteric inhibition constant.

Allosteric Activation

Conversely, allosteric modulation can also activate enzymes. The allosteric activation can be described using a similar equation, with appropriate modifications based on the nature of activation.

Applications in Drug Discovery

Understanding allosteric modulation provides opportunities for drug discovery. Modulating enzyme activity allosterically allows for the development of selective and targeted therapeutics. Case studies of successful drugs leveraging allosteric modulation can offer valuable insights.

Challenges and Future Directions

Despite the promise of allosteric modulation, challenges include predicting allosteric sites and designing molecules with desired modulatory effects. Future directions involve integrating structural biology, computational modeling, and experimental validation to advance our understanding of allosteric mechanisms.

Exploring allosteric modulation in enzymes opens avenues for the development of innovative drugs with enhanced specificity and fewer side effects.

7.3.11   In Silico Design of Metalloenzyme Inhibitors

Metalloenzymes play crucial roles in various biological processes, and their dysregulation is often associated with diseases. In this subsection, we delve into the in silico design strategies for developing inhibitors targeting metalloenzymes.

Role of Metalloenzymes

Metalloenzymes, which contain metal ions as essential cofactors, participate in catalyzing key biochemical reactions. Examples include zinc-dependent metalloproteases and copper-containing oxidases. Inhibiting these enzymes can be a viable therapeutic strategy.

Computational Methods for Inhibitor Design

In silico design of metalloenzyme inhibitors involves employing computational methods to predict, analyze, and optimize potential inhibitors. Molecular docking, molecular dynamics simulations, and quantum mechanical calculations are integral tools in this process.

Mathematical Models for Binding Affinity

The binding affinity between a metalloenzyme and its inhibitor can be represented mathematically. The scoring functions used in molecular docking often involve terms related to energy contributions, such as van der Waals forces, electrostatic interactions, and metal-ligand coordination.

Metal Coordination Geometry

Understanding the metal coordination geometry is crucial for designing effective metalloenzyme inhibitors. In silico methods enable the exploration of various coordination patterns and help predict the stability of metal-inhibitor complexes.

Quantum Mechanical Approaches

In some cases, a more detailed understanding of metal-ligand interactions requires quantum mechanical calculations. Density functional theory (DFT) and ab initio methods can provide insights into electronic structure and energetics.

Case Studies

Several successful inhibitors of metalloenzymes have been developed using in silico approaches. Case studies highlight the application of these methods in designing molecules that selectively target the metalloenzyme active sites.

Challenges and Considerations

Designing metalloenzyme inhibitors poses challenges due to the intricacies of metal-ligand interactions. Overcoming challenges such as metal selectivity, off-target effects, and predicting in vivo efficacy remains an ongoing area of research.

Future Perspectives

Advancements in computational power and methodologies continue to enhance the accuracy of in silico design. Integrating experimental data and leveraging machine learning approaches hold promise for more effective and precise metalloenzyme inhibitor discovery.

In silico design of metalloenzyme inhibitors represents a valuable approach in the quest for novel therapeutics, offering insights into the complex interplay between metal ions and biological macromolecules.

7.3.12 Drug-Drug Interaction Studies: Cytochrome P450 Binding

Cytochrome P450 (CYP) enzymes play a pivotal role in drug metabolism, and drug-drug interactions (DDIs) involving CYP binding can significantly impact the pharmacokinetics of co-administered drugs. This subsection explores computational approaches to study DDIs, focusing on CYP binding.

Role of Cytochrome P450 in Drug Metabolism

Cytochrome P450 enzymes, found predominantly in the liver, are responsible for metabolizing a wide range of drugs. Understanding how drugs interact with these enzymes is crucial for predicting potential DDIs.

Computational Modeling of Cytochrome P450 Binding

In silico methods provide a powerful means to study drug binding to CYP enzymes. Molecular docking, molecular dynamics simulations, and quantum mechanical calculations contribute to the computational modeling of drug-CYP interactions.

Mathematical Representation of Drug Binding

Mathematical models are employed to represent the binding affinity and kinetics of drugs interacting with CYP enzymes. These models often incorporate parameters related to ligand-receptor interactions, such as dissociation constants and reaction rates.

Structural Insights into Drug-Drug Interactions

Computational studies yield structural insights into the binding modes of drugs within the CYP active site. Visualization of these interactions helps in understanding the mechanisms underlying DDIs and predicting their potential impact.

Impact on Drug Pharmacokinetics

By studying the binding of multiple drugs to CYP enzymes, researchers can predict potential DDIs and assess their impact on drug pharmacokinetics. Computational tools aid in quantifying the likelihood and severity of interactions.

Quantitative Structure-Activity Relationships (QSAR)

QSAR models are employed to establish relationships between the chemical structures of drugs and their potential to interact with CYP enzymes. These quantitative models assist in predicting DDIs based on structural features.

Case Studies in Cytochrome P450-Related DDIs

Real-world case studies highlight instances where DDIs involving CYP binding have been successfully predicted and validated. These examples demonstrate the utility of computational methods in anticipating and managing drug interactions.

Challenges and Considerations

Despite advancements, challenges remain in accurately predicting the complex landscape of DDIs. Factors such as substrate promiscuity, genetic polymorphisms, and the interplay of multiple enzymes contribute to the complexity of these interactions.

Future Directions

Improving the accuracy of computational models, incorporating personalized medicine approaches, and integrating experimental data are crucial for advancing the field of drug-drug interaction studies. Future research aims to enhance the prediction and understanding of DDIs involving cytochrome P450.

The computational exploration of drug-drug interactions, particularly in the context of cytochrome P450 binding, contributes valuable insights to drug development and clinical practice.

7.3.13 Prediction of Drug Metabolism: CYP2D6 Substrate Specificity

The prediction of drug metabolism is a critical aspect of drug development, and understanding the substrate specificity of cytochrome P450 (CYP) enzymes, such as CYP2D6, is essential for anticipating metabolic outcomes.

Role of CYP2D6 in Drug Metabolism

CYP2D6 is a major drug-metabolizing enzyme responsible for the biotransformation of a diverse array of drugs. Its substrate specificity varies widely, and predicting which drugs are metabolized by CYP2D6 is crucial for assessing potential drug-drug interactions.

Computational Methods for Predicting CYP2D6 Substrate Specificity

Various computational approaches are employed to predict the substrate specificity of CYP2D6. Machine learning models, molecular docking simulations, and quantitative structure-activity relationship (QSAR) analyses are among the methods used to elucidate the interaction patterns between drugs and CYP2D6.

Machine Learning Models

Machine learning algorithms, trained on experimental data, can predict whether a given compound is likely to be metabolized by CYP2D6. These models utilize features such as molecular descriptors, chemical fingerprints, and structural properties to make predictions.

Molecular Docking Simulations

Molecular docking simulations enable the exploration of the binding modes and affinities of drugs with CYP2D6. These simulations provide insights into the three-dimensional interactions between substrates and the enzyme's active site.

Quantitative Structure-Activity Relationship (QSAR)

QSAR models establish correlations between the chemical structure of drugs and their likelihood of being metabolized by CYP2D6. These quantitative models are valuable for predicting substrate specificity based on structural features.

Experimental Validation of Predictions

Predictions made by computational models are often experimentally validated to confirm their accuracy. This involves conducting in vitro and in vivo studies to assess the actual metabolism of drugs by CYP2D6, validating the computational predictions.

Case Studies: Successful Prediction of CYP2D6 Substrates

Real-world case studies demonstrate instances where computational models successfully predicted CYP2D6 substrate specificity. These examples showcase the practical applications of predictive models in drug development and metabolism studies.

Challenges and Considerations

Despite advancements, challenges exist in accurately predicting CYP2D6 substrate specificity due to the enzyme's complex substrate promiscuity and the influence of genetic polymorphisms. Addressing these challenges is crucial for improving the reliability of predictions.

Future Perspectives

The future of predicting drug metabolism involving CYP2D6 substrate specificity lies in the refinement of computational models, integration with high-throughput screening data, and advancements in understanding the intricate mechanisms of CYP2D6-mediated metabolism.

Understanding and predicting CYP2D6 substrate specificity contribute to more informed drug development processes and facilitate the identification of potential drug interactions and metabolic pathways.

7.3.14 Designing Antifungal Agents: Targeting Ergosterol Synthesis

Fungal infections pose a significant threat to human health, and the development of antifungal agents targeting essential pathways in fungi, such as ergosterol synthesis, is a key area of research.

Importance of Ergosterol in Fungal Cells

Ergosterol is a vital component of fungal cell membranes, playing a role similar to cholesterol in animal cells. Disrupting ergosterol synthesis can lead to membrane instability and cell death, making it an attractive target for antifungal drug design.

Inhibition of Ergosterol Biosynthetic Enzymes

Antifungal agents designed to target ergosterol synthesis often focus on inhibiting key enzymes involved in the biosynthetic pathway. These enzymes include lanosterol 14α-demethylase, 24-methylene dihydroxycholesterol reductase, and others.

Molecular Docking Studies

Molecular docking studies are employed to understand the interaction between potential antifungal compounds and ergosterol biosynthetic enzymes. These computational simulations aid in predicting the binding affinity and mode of action of designed inhibitors.

Structure-Based Drug Design

Structure-based drug design techniques utilize the three-dimensional structures of ergosterol biosynthetic enzymes to rationally design molecules that can bind to these targets with high specificity. This approach enhances the likelihood of developing potent antifungal agents.

In Vivo Efficacy Studies

Promising antifungal agents identified through computational and in vitro studies undergo in vivo efficacy evaluations. These studies involve testing the designed compounds in animal models of fungal infections to assess their therapeutic potential.

Synergy with Existing Antifungal Drugs

Combining novel ergosterol synthesis inhibitors with existing antifungal drugs is explored to enhance treatment efficacy and mitigate the risk of resistance. Synergistic effects can lead to improved therapeutic outcomes.

Challenges in Antifungal Drug Design

Designing effective antifungal agents targeting ergosterol synthesis faces challenges, including the need for selectivity against human sterol biosynthesis and addressing the emergence of resistance. Overcoming these challenges is crucial for developing successful antifungal therapies.

Case Studies: Successful Design of Ergosterol Synthesis Inhibitors

Several case studies highlight successful instances where antifungal agents designed to target ergosterol synthesis have demonstrated efficacy in preclinical and clinical settings. These examples showcase the translational potential of such drug design strategies.

Future Directions

Future directions in designing antifungal agents involve exploring innovative approaches, leveraging advanced technologies, and incorporating a deeper understanding of fungal biology. Continued research in this area is essential for addressing emerging fungal threats and improving treatment options.

Designing antifungal agents targeting ergosterol synthesis remains a dynamic and evolving field with the potential to significantly impact the treatment of fungal infections.

7.3.15 Optimizing Anti-HIV Protease Inhibitors

Human Immunodeficiency Virus (HIV) protease is a critical enzyme for viral replication, and developing effective inhibitors is crucial for antiretroviral therapy. Optimization of anti-HIV protease inhibitors involves a multidisciplinary approach integrating computational modeling, medicinal chemistry, and biological assays.

Understanding HIV Protease and Its Role

HIV protease plays a key role in the final stages of the viral life cycle, cleaving viral polyproteins into functional proteins necessary for the assembly of new virions. Inhibiting this enzyme disrupts viral maturation, making it an attractive target for drug development.

Computational Modeling of Protease Inhibitors

Molecular docking and dynamics simulations are employed to model the interaction between potential inhibitors and the active site of HIV protease. These computational approaches provide insights into binding affinities, binding modes, and structural features crucial for inhibitor design.

Structure-Activity Relationship (SAR) Studies

Optimization begins with the exploration of the structure-activity relationship of existing protease inhibitors. Medicinal chemists analyze the impact of chemical modifications on the inhibitory potency, bioavailability, and other pharmacokinetic properties.

De Novo Design and Virtual Screening

De novo design and virtual screening are utilized to generate novel anti-HIV protease inhibitors. Computational algorithms predict the chemical structures that are likely to have favorable interactions with the protease active site, guiding the synthesis of new compounds.

Peptidomimetic Design Strategies

Many HIV protease inhibitors mimic the natural substrates of the enzyme. Peptidomimetic design involves creating synthetic molecules that resemble the structure of the natural substrate, optimizing them for increased potency and pharmacokinetic properties.

In Vitro and In Vivo Evaluation

Promising anti-HIV protease inhibitors undergo rigorous in vitro testing using biochemical assays and cell-based systems. Compounds demonstrating efficacy and minimal cytotoxicity proceed to in vivo evaluations using animal models, assessing pharmacokinetics and antiviral activity.

Resistance Studies and Combination Therapies

Anticipating the development of resistance, researchers conduct resistance studies to understand the potential mutations that may arise. Optimization may involve the development of combination therapies to target multiple sites in the viral life cycle, reducing the likelihood of resistance.

Clinical Trials and Regulatory Approval

Compounds showing favorable results in preclinical studies advance to clinical trials. Clinical trials assess safety, efficacy, and dosing regimens in human subjects. Successful trials lead to regulatory approval, allowing the inhibitor to become part of antiretroviral therapy.

Patient-Specific Approaches

Tailoring anti-HIV protease inhibitors to individual patient profiles is an emerging area. Personalized medicine approaches consider genetic variations in both the virus and the host, optimizing treatment outcomes while minimizing side effects.

Challenges and Future Perspectives

Optimizing anti-HIV protease inhibitors faces challenges such as drug resistance, side effects, and the need for continuous innovation. Future perspectives involve the development of next-generation inhibitors, improved drug delivery methods, and advancing towards a functional cure for HIV.

Case Studies: Successful Anti-HIV Protease Inhibitors

Several case studies highlight the success stories of anti-HIV protease inhibitors, emphasizing their impact on improving the quality of life for individuals living with HIV. These examples showcase the translational potential of optimized protease inhibitors.

7.3.16 Understanding Protein-Nucleic Acid Interactions: DNA Binding Proteins

DNA binding proteins play a crucial role in various cellular processes, including gene regulation, DNA replication, and repair. Understanding the interactions between proteins and DNA is fundamental for unraveling the molecular mechanisms governing these processes.

Importance of DNA Binding Proteins

DNA binding proteins are diverse in function and structure, encompassing transcription factors, DNA repair enzymes, and chromatin remodeling proteins. Their ability to specifically recognize and bind to DNA sequences is pivotal for orchestrating cellular activities.

Structural Basis of Protein-DNA Interactions

The structural basis of protein-DNA interactions involves the recognition of specific DNA sequences by protein domains. Major groove and minor groove interactions, hydrogen bonding, and van der Waals forces contribute to the formation of stable complexes between DNA and binding proteins.

DNA Recognition Motifs

Proteins that interact with DNA often contain specific DNA recognition motifs, such as helix-turn-helix, zinc finger, and leucine zipper motifs. These motifs enable proteins to recognize and bind to distinct DNA sequences with high affinity and specificity.

Role of Electrostatics and Hydrogen Bonding

Electrostatic interactions and hydrogen bonding between amino acid residues of the protein and nucleotide bases of DNA contribute to the specificity of binding. Positively charged residues often interact with the negatively charged phosphate backbone of DNA.

Dynamic Nature of Protein-DNA Complexes

Protein-DNA interactions are dynamic, involving conformational changes in both the protein and DNA. These dynamic complexes allow for processes such as DNA bending, looping, and unwinding, crucial for various cellular functions.

Experimental Techniques for Studying Protein-DNA Interactions

Various experimental techniques are employed to study protein-DNA interactions, including X-ray crystallography, nuclear magnetic resonance (NMR) spectroscopy, and electrophoretic mobility shift assays (EMSA). These techniques provide insights into the structure, dynamics, and binding affinity of protein-DNA complexes.

Computational Approaches to Predicting Binding Sites

Computational methods, such as molecular docking and molecular dynamics simulations, are utilized to predict and analyze protein-DNA binding sites. These approaches aid in understanding the energetics and kinetics of interactions and can guide experimental studies.

Biological Significance of Protein-DNA Interactions

Protein-DNA interactions have profound biological significance, influencing gene expression, DNA replication fidelity, and repair mechanisms. Dysregulation of these interactions is associated with various diseases, including cancer and genetic disorders.

Engineering DNA Binding Proteins

The engineering of DNA binding proteins involves modifying existing proteins or designing novel ones with tailored DNA binding specificities. This field has applications in synthetic biology, gene therapy, and the development of tools for genome editing.

Challenges and Future Directions

Despite significant progress, challenges in understanding the complexity of protein-DNA interactions persist. Future directions involve integrating multi-omics data, exploring protein dynamics in live cells, and developing therapeutic interventions targeting aberrant interactions.

Case Studies: DNA Binding Proteins in Action

Several case studies highlight the functional roles of DNA binding proteins in specific cellular processes. These examples illustrate the diverse functions and mechanisms through which DNA binding proteins contribute to cellular homeostasis.

7.3.17   Advancements in Drug Delivery: Designing Nanocarriers

Nanocarriers have emerged as promising vehicles for drug delivery, offering improved therapeutic outcomes and reduced side effects. The design of nanocarriers involves intricate formulations to ensure efficient drug encapsulation, targeted delivery, and controlled release.

Nanoformulations for Drug Delivery

Nanoformulations encompass a variety of nanocarriers, including liposomes, micelles, nanoparticles, and dendrimers. These carriers are designed to encapsulate drugs, protect them from degradation, and deliver them to specific target sites within the body.

Mathematical Formulas for Drug Encapsulation Efficiency

The encapsulation efficiency (EE) of a nanocarrier, representing the percentage of drug encapsulated within the carrier, can be mathematically expressed as:

$$\text{Encapsulation Efficiency (EE)} = \frac{\text{Amount of Encapsulated Drug}}{\text{Total Amount of Drug}} \times 100\%$$
$$(7.20)$$

Efficient drug encapsulation ensures maximum therapeutic benefit and minimizes drug wastage.

Chemical Formulas for Nanocarrier Components

The composition of nanocarriers involves various components, such as lipids, polymers, and surfactants. The chemical formula for a liposome, a commonly used nanocarrier, can be represented as:

$$\text{Liposome: } (\text{Lipid})_n \qquad\qquad (7.21)$$

Here, n represents the number of lipid molecules forming the liposome.

Targeted Drug Delivery Strategies

Nanocarriers enable targeted drug delivery by functionalizing their surfaces with ligands that can recognize and bind to specific receptors on target cells. The targeted drug delivery efficiency (TDE) can be expressed as:

$$TDE = \frac{\text{Amount of Drug Delivered to Target Site}}{\text{Total Amount of Administered Drug}} \times 100\% \qquad (7.22)$$

Controlled Release Kinetics

The release of drugs from nanocarriers follows controlled kinetics. The mathematical expression for drug release over time (t) can be described by the Higuchi model:

$$\text{Drug Release} = k\sqrt{t} \qquad (7.23)$$

Where k is the release rate constant.

Design Considerations for Nanocarriers

The design of nanocarriers involves considerations such as particle size, surface charge, and stability. The mathematical relationship between particle size (D) and drug diffusion coefficient (D_0) is given by the Stokes-Einstein equation:

$$D = \frac{k_B T}{3\pi \eta D_0} \qquad (7.24)$$

Where k_B is the Boltzmann constant, T is the absolute temperature, and η is the viscosity.

Biocompatibility and Toxicity Assessments

Ensuring the biocompatibility of nanocarriers is essential. The toxicity index (TI) can be calculated as:

$$TI = \frac{\text{LD50 (Dose Inducing Toxicity)}}{\text{ED50 (Effective Dose)}} \qquad (7.25)$$

Case Studies: Successful Nanocarrier Applications

Several case studies showcase the success of nanocarrier-based drug delivery in treating various diseases. These examples highlight the versatility and potential of nanocarriers in improving therapeutic outcomes.

7.3.18   Integration of Machine Learning in Drug Design

The integration of machine learning (ML) techniques in drug design has revolutionized the process of identifying potential drug candidates. ML models leverage large datasets to predict molecular properties, interactions, and bioactivities, aiding in the rational design of novel therapeutics.

Machine Learning Algorithms in Drug Design

Various ML algorithms are employed in drug design, including:

- **Random Forests:** A decision tree ensemble method capable of handling complex relationships in data.

- **Support Vector Machines (SVM):** Useful for classification and regression tasks, SVMs are applied to predict bioactivity and drug-likeness.

- **Neural Networks:** Deep learning models, such as convolutional neural networks (CNNs) and recurrent neural networks (RNNs), demonstrate proficiency in feature learning and molecular representation.

- **Generative Models:** Models like variational autoencoders (VAEs) and generative adversarial networks (GANs) assist in generating novel molecular structures.

Mathematical Formulas for ML Model Training

The training of ML models involves minimizing a loss function. For a generic ML model with parameters θ, the optimization problem is formulated as:

$$\min_{\theta} \mathcal{L}(\theta) \tag{7.26}$$

Here, $\mathcal{L}(\theta)$ represents the loss function, which quantifies the difference between predicted and actual outcomes.

Chemoinformatics Features and Descriptors

Chemoinformatics plays a crucial role in extracting molecular features for ML models. Descriptors, such as molecular fingerprints, physicochemical properties, and structural motifs, are mathematically represented as vectors:

$$\text{Molecular Descriptor Vector: } \mathbf{X} = [x_1, x_2, \ldots, x_n] \tag{7.27}$$

Validation Metrics for ML Models

The performance of ML models is assessed using various metrics, including:

- **Accuracy** (ACC): The proportion of correctly predicted instances.

- **Precision** (PR): The ratio of true positive predictions to the total predicted positives.

- **Recall** (RE): The ratio of true positive predictions to the total actual positives.

- **F1 Score** ($F1$): The harmonic mean of precision and recall.

These metrics quantify the model's ability to make accurate predictions.

Case Studies: Successful ML Applications in Drug Design

Numerous case studies demonstrate the success of ML in drug design, from predicting bioactivity to optimizing molecular structures. These applications showcase the versatility and efficiency of ML in expediting drug discovery processes.

Chapter 8

Bioinformatics in Molecular Docking

8.1 Integration with Bioinformatics Tools

Molecular docking, at the intersection of computational biology and bioinformatics, benefits significantly from the integration with various bioinformatics tools. This section explores the synergies between molecular docking and bioinformatics, elucidating how computational techniques and data analysis tools enhance the accuracy and scope of docking studies.

8.1.1 Bioinformatics Databases and Molecular Docking

Bioinformatics databases play a pivotal role in molecular docking studies by providing a wealth of biological data. Integration with databases such as the Protein Data Bank (PDB) allows researchers to access experimentally determined structures of biomolecules, serving as essential inputs for docking simulations. The utilization of bioinformatics tools for data curation ensures the accuracy and reliability of structures used in docking experiments.

8.1.2   Structural Bioinformatics for Target Identification

Structural bioinformatics tools aid in the identification of suitable targets for molecular docking studies. Through the analysis of protein structures and their binding sites, researchers can prioritize potential targets based on druggability and relevance to specific diseases. This integration enhances the rational selection of proteins for docking experiments, streamlining the drug discovery process.

8.1.3   Molecular Dynamics Simulations and Docking

The integration of molecular dynamics (MD) simulations with molecular docking provides a dynamic perspective of ligand-receptor interactions. MD simulations offer insights into the flexibility and conformational changes of biomolecules over time, complementing the static snapshots provided by docking studies. By integrating MD trajectories, researchers gain a more comprehensive understanding of ligand binding and unbinding events.

8.1.4   Bioinformatics Tools for Ligand Screening

Bioinformatics tools contribute to ligand screening strategies in molecular docking. Virtual screening pipelines often incorporate ligand-based bioinformatics approaches, such as quantitative structure-activity relationship (QSAR) modeling and pharmacophore analysis. These tools assist in the selection of diverse ligands with optimal chemical properties, enriching the pool of compounds subjected to docking studies.

8.1.5   Pathway Analysis and Systems Biology Integration

The integration of molecular docking with pathway analysis tools enhances the understanding of biological processes at a systems level. Bioinformatics tools enable the mapping of molecular interactions within cellular pathways, elucidating the impact of ligand binding on signaling cascades. This systems biology approach guides the interpretation of docking results in the context of broader cellular functions.

8.1.6 Genomic Data for Personalized Medicine

In the era of personalized medicine, the integration of genomic data with molecular docking is gaining prominence. Bioinformatics tools analyze patient-specific genomic information to identify genetic variations that may influence drug responses. Molecular docking studies then consider these variations in predicting individualized drug binding affinities, contributing to personalized therapeutic strategies.

8.1.7 Integrating Network Pharmacology in Docking Studies

Network pharmacology, a bioinformatics-driven approach, integrates molecular docking with network analysis to explore the interactions between drugs and multiple targets within biological networks. This holistic perspective enables the identification of polypharmacological effects and off-target interactions, guiding the design of more effective and safer drug candidates.

8.1.8 Metabolomics and Docking: Unraveling Metabolic Pathways

The integration of metabolomics data with molecular docking provides insights into metabolic pathways and their impact on drug metabolism. Bioinformatics tools analyze metabolite profiles to identify potential interactions between drugs and endogenous metabolites. Understanding these interactions aids in predicting drug metabolism and potential side effects.

8.1.9   Machine Learning for Predictive Docking Models

Machine learning algorithms, a subset of bioinformatics tools, contribute to the development of predictive docking models. By training on large datasets of experimentally validated binding affinities, machine learning models enhance the accuracy of scoring functions used in molecular docking. This integration accelerates the identification of lead compounds with high binding potentials.

8.1.10   Bioinformatics Approaches for Solvent Effects

Incorporating bioinformatics approaches to model solvent effects refines the accuracy of molecular docking studies. Solvent molecules play a crucial role in ligand-receptor interactions, impacting binding affinities. Bioinformatics tools simulate the solvent environment, allowing researchers to account for its influence on the energetics of ligand binding.

8.1.11   Data Integration for Systems Pharmacology

Systems pharmacology, an interdisciplinary field encompassing bioinformatics, integrates diverse omics data with molecular docking. By combining genomics, proteomics, and metabolomics data, researchers construct comprehensive models of drug action. This data-driven approach enhances the prediction of drug effects, facilitating the design of multi-targeted therapies.

8.1.12   Examples of Bioinformatics-Integrated Docking Studies

1. **Targeting Oncogenic Mutations:** Integrating bioinformatics tools to analyze cancer genomics data facilitates the identification of oncogenic mutations. Docking studies then prioritize small molecules that specifically target these mutated proteins, offering potential therapies for personalized cancer treatment.

   2. **Drug-Drug Interaction Prediction:** Bioinformatics tools predict potential drug-drug interactions by analyzing the chemical structures and phar-

macological profiles of multiple drugs. Molecular docking studies validate these predictions, elucidating the mechanisms and impacts of co-administered drugs.

3. **Structural Bioinformatics in Antiviral Drug Design:** Analyzing the structures of viral proteins using structural bioinformatics tools aids in the design of antiviral drugs. Molecular docking studies assess the binding affinities of potential inhibitors, guiding the development of effective antiviral therapies.

8.1.13 Challenges and Future Directions

Despite the advantages of integrating molecular docking with bioinformatics tools, challenges persist. Standardization of data formats, interoperability between diverse tools, and the need for advanced algorithms pose ongoing challenges. Future directions involve developing unified platforms that seamlessly integrate molecular docking with a broader array of bioinformatics approaches, fostering a more collaborative and efficient drug discovery process.

8.2 Structural Bioinformatics

Structural bioinformatics is a pivotal component in the field of molecular docking, providing essential tools and methodologies for understanding the three-dimensional structures of biomolecules. This section delves into the role of structural bioinformatics in molecular docking studies, elucidating its significance in target identification, ligand screening, and the rational design of novel therapeutic agents.

8.2.1 Protein Structure Determination

Protein structure determination is a critical step in understanding the function and interactions of biological macromolecules. Various experimental and computational techniques are employed to elucidate the three-dimensional arrangement of atoms in proteins.

X-ray Crystallography

X-ray crystallography is a widely used experimental method for determining protein structures. The process involves crystallizing the protein and exposing it to X-rays. The resulting diffraction pattern is analyzed to reconstruct the electron density map of the protein, which is then used to determine the atomic coordinates.

Nuclear Magnetic Resonance (NMR) Spectroscopy

NMR spectroscopy is another powerful experimental technique for protein structure determination. In this method, proteins are studied in solution, and the interaction of nuclear spins with magnetic fields provides information about atomic distances and angles. NMR is particularly valuable for studying dynamic proteins.

Cryo-Electron Microscopy (Cryo-EM)

Cryo-EM has emerged as a revolutionary technique for visualizing large macromolecular complexes at near-atomic resolution. In this method, protein samples are rapidly frozen, and electron micrographs of the frozen samples are used to reconstruct 3D structures. Cryo-EM is especially useful for studying flexible and heterogeneous structures.

Computational Modeling

Computational methods, such as molecular dynamics simulations and homology modeling, play a vital role in predicting and refining protein structures. Molecular dynamics simulations simulate the motion of atoms over time, providing insights into protein dynamics. Homology modeling leverages known protein structures to predict the structure of a related protein.

Mathematical Formulas in Structure Determination

The mathematical formulas involved in structure determination often include the analysis of experimental data and the optimization of atomic coordinates. For example, in X-ray crystallography, the Fourier transform is used to convert diffraction data into an electron density map.

Validation Metrics for Protein Structures

Validation metrics ensure the reliability of experimentally determined or computationally predicted protein structures. Common metrics include:

- **R-factor:** Measures the agreement between observed and calculated structure factors in X-ray crystallography.

- **RMSD (Root Mean Square Deviation):** Quantifies the average distance between corresponding atoms in a superimposed structure.

- **MolProbity Score:** Evaluates the stereochemical quality of a protein structure.

These metrics aid researchers in assessing the accuracy and precision of protein structures.

8.2.2 Homology Modeling for Target Prediction

Homology modeling, also known as comparative modeling, is a computational technique used to predict the three-dimensional structure of a target protein based on the known structure of a homologous template protein. This method is particularly valuable when experimental structures for the target protein are not available.

Procedure for Homology Modeling

The homology modeling process typically involves the following steps:

1. **Template Selection:** Choose a template protein with a known structure that shares significant sequence similarity with the target protein.

2. **Sequence Alignment:** Align the target protein's amino acid sequence with the template protein's sequence, ensuring proper alignment of corresponding residues.

3. **Model Building:** Generate a three-dimensional model of the target protein based on the aligned sequences. This involves transferring the spatial coordinates of atoms from the template to the target.

4. **Model Refinement:** Refine the initial model through energy minimization, optimizing the geometry and removing steric clashes.

5. **Validation:** Assess the quality of the homology model using validation tools, ensuring that it conforms to stereochemical and experimental constraints.

Mathematical Formulas in Homology Modeling

The mathematical formulas involved in homology modeling often include scoring functions for energy minimization and validation metrics. For example, energy minimization may use force field equations to optimize the spatial arrangement of atoms, and validation metrics such as Ramachandran plots quantify the model's stereochemical quality.

Applications in Target Prediction

Homology modeling is widely employed in drug discovery for predicting the structures of target proteins relevant to diseases. It aids in understanding the binding sites, interactions, and conformational changes of target proteins, facilitating the design of specific ligands.

Challenges and Considerations

Despite its utility, homology modeling comes with challenges, such as accuracy limitations in regions of low sequence similarity and potential errors in loop modeling. Researchers must carefully evaluate the reliability of homology models and consider alternative methods when necessary.

8.2.3 Molecular Docking with Experimental Structures

Molecular docking is a powerful computational technique that predicts the preferred orientation of one molecule (the ligand) when bound to another molecule (the receptor) to form a stable complex. When experimental structures of the target proteins are available, molecular docking can be applied to gain insights into ligand-receptor interactions and predict binding affinities.

Utilizing Experimental Structures

The integration of experimental structures into molecular docking involves the following key steps:

1. **Structure Preparation:** Obtain the experimental structure of the target protein and preprocess it to ensure proper geometry and removal of water molecules.

2. **Ligand Preparation:** Prepare ligands by optimizing their geometry, assigning charges, and adding hydrogen atoms. This step ensures that ligands are in a suitable conformation for docking.

3. **Docking Parameters:** Set parameters for the docking algorithm, specifying search algorithms, scoring functions, and other relevant parameters.

4. **Docking Simulations:** Perform docking simulations by exploring possible binding modes of ligands within the binding site of the target protein.

5. **Scoring and Analysis:** Evaluate and score the generated poses based on energy calculations and other scoring functions. Analyze the results to

identify potential binding modes.

Mathematical Formulas in Docking

Molecular docking involves several mathematical formulas, including scoring functions to estimate binding affinity. One commonly used scoring function is the empirical free energy scoring equation:

$$\text{Binding Free Energy} = \text{Internal Energy} + \text{Ligand-Receptor Interaction Energy} - \text{Entropy Contribution}$$

$$(8.1)$$

Here, the internal energy represents the energy of the ligand and receptor in isolation, and the ligand-receptor interaction energy accounts for the intermolecular forces upon binding.

Applications and Advantages

Molecular docking with experimental structures finds applications in rational drug design, lead optimization, and understanding ligand-binding mechanisms. The use of experimental structures enhances the accuracy of predictions and provides a more realistic representation of the ligand-receptor complex.

Considerations and Challenges

Despite its advantages, molecular docking with experimental structures faces challenges related to flexibility, solvent effects, and the accurate representation of binding events. Researchers must consider these factors when interpreting docking results and validating predictions.

8.2.4   Structural Bioinformatics in Virtual Screening

Structural bioinformatics plays a crucial role in the success of virtual screening, a computational method used in drug discovery to identify potential ligands with high binding affinities for a target protein. By leveraging structural information and bioinformatics tools, virtual screening enhances the efficiency of identifying lead compounds.

Integration of Structural Information

The success of virtual screening heavily relies on the availability and accuracy of structural information for both the target protein and the screened ligands. Structural bioinformatics tools are employed to process, analyze, and visualize these structures.

Mathematical Formulas in Virtual Screening

Mathematical formulas are integral to the scoring functions used in virtual screening algorithms. A commonly employed scoring function is based on the molecular docking score, which combines terms representing van der Waals forces, electrostatic interactions, and other factors:

$$\text{Docking Score} = \text{Van der Waals Term} + \text{Electrostatic Term} + \dots \quad (8.2)$$

Here, the terms quantify the favorable and unfavorable interactions between the ligand and the target protein.

Bioinformatics Tools for Ligand Screening

Several bioinformatics tools are employed in virtual screening to enhance the analysis of ligands and improve the accuracy of predictions. These tools include:

- **Structure-Based Virtual Screening Tools:** Identify potential ligands based on their structural complementarity with the target protein.

- **Pharmacophore Modeling:** Define the essential features of ligands that are crucial for binding, aiding in the screening of compound databases.

- **Molecular Dynamics Simulations:** Explore the dynamic behavior of ligand-receptor interactions to refine virtual screening predictions.

Applications and Case Studies

Structural bioinformatics in virtual screening has been successfully applied in various drug discovery projects. Case studies demonstrating the integration of structural bioinformatics tools in virtual screening processes provide insights into the identification of potential drug candidates.

Challenges and Future Directions

Despite advancements, challenges remain, such as the accurate representation of protein flexibility and the consideration of solvent effects. Future directions in structural bioinformatics for virtual screening involve addressing these challenges and further improving the predictive capabilities of virtual screening methods.

8.2.5   Protein-Ligand Interaction Analysis

Protein-ligand interaction analysis is a critical step in understanding the binding affinity and specificity of ligands to target proteins. Various methods and mathematical formulations are employed to analyze these interactions.

Mathematical Formulas for Binding Affinity

The binding affinity (B_{affinity}) between a protein (P) and a ligand (L) can be expressed using the following equation:

$$B_{\text{affinity}} = \frac{1}{K_d} = \frac{[\text{PL}]}{[\text{P}][\text{L}]} \tag{8.3}$$

Here, K_d represents the dissociation constant, and [PL], [P], and [L] are the concentrations of the protein-ligand complex, free protein, and free ligand, respectively.

Chemical Formulas for Binding Interactions

Chemical formulas can represent specific types of interactions between proteins and ligands. For instance, hydrogen bonding can be expressed as:

$$P-O\cdots H-L \tag{8.4}$$

This notation illustrates a hydrogen bond between the protein (P) and the ligand (L).

Energy Contributions in Protein-Ligand Interactions

The total binding energy (E_{bind}) can be decomposed into different energy contributions:

$$E_{bind} = E_{vdW} + E_{ele} + E_{hbond} + \cdots \tag{8.5}$$

Here, E_{vdW} represents van der Waals interactions, E_{ele} represents electrostatic interactions, E_{hbond} represents hydrogen bonding interactions, and so on.

Visualization with Molecular Graphics

Molecular graphics tools are commonly used to visually represent protein-ligand interactions. Diagrams can illustrate key interactions, such as hydrophobic contacts, pi-stacking, and specific residue interactions.

Quantifying Specificity and Selectivity

Specificity and selectivity of ligands can be quantified using metrics such as the selectivity score (S_{score}):

$$S_{score} = \frac{\text{Affinity for Target A}}{\text{Affinity for Target B}} \tag{8.6}$$

A high selectivity score indicates preferential binding to a specific target.

8.2.6 Pharmacophore Modeling for Ligand Design

Pharmacophore modeling is a crucial approach in ligand design, aiming to identify the essential structural and chemical features that contribute to a ligand's binding affinity for a target receptor.

Mathematical Formulation

The pharmacophore model is often represented as a mathematical formula, where ϕ denotes the pharmacophoric features:

$$\text{Pharmacophore Model:} \quad \phi_1 \cap \phi_2 \cap \ldots \cap \phi_n \tag{8.7}$$

Here, $\cap$ represents the intersection, and ϕ_i denotes each pharmacophoric feature.

Chemical Features in Pharmacophore Models

Pharmacophoric features include elements such as hydrogen bond donors, hydrogen bond acceptors, hydrophobic regions, and aromatic rings. These can be represented using chemical formulas, e.g.:

$$\text{HBD:} \quad \text{P}-\text{H}\cdots\text{L} \tag{8.8}$$

This represents a hydrogen bond donor (HBD) interaction between a pharmacophoric feature on the ligand (L) and the protein (P).

Geometric Constraints

In pharmacophore modeling, geometric constraints are crucial for accurately representing the spatial arrangement of pharmacophoric features. This can be expressed mathematically as:

$$\text{Distance Constraint:} \quad d_{\text{H-bond}} < 3 \text{ Å} \tag{8.9}$$

This constraint ensures that the distance ($d_{\text{H-bond}}$) between a hydrogen bond donor and acceptor is less than 3 angstroms.

Validation Metrics

Pharmacophore models are validated using various metrics, such as the cost function, which quantifies the agreement between experimental and predicted features:

$$\text{Cost Function:} \quad \text{Cost} = \sum_i w_i (E_i - O_i)^2 \tag{8.10}$$

Here, w_i represents the weight, E_i is the predicted value, and O_i is the observed value for each feature.

Application in Ligand Design

Pharmacophore modeling guides the design of ligands with improved binding affinity by considering the essential features required for interaction with the target receptor.

8.2.7 Integration of Protein Flexibility

Integrating protein flexibility into molecular docking simulations is essential for a more accurate representation of ligand-receptor interactions. Traditional docking approaches often assume rigid protein structures, which may lead to incomplete insights into the binding process.

Mathematical Formulation

Incorporating protein flexibility can be mathematically expressed through the inclusion of terms representing the dynamic nature of the protein-ligand complex:

$$\text{Total Energy} = \text{Internal Energy} + \text{Interaction Energy} + \text{Entropy} \tag{8.11}$$

Here, the internal energy accounts for the protein's conformational changes, the interaction energy represents the binding affinity, and the entropy term considers the disorder in the system.

Flexibility Models

Different models capture protein flexibility, such as:

- **Normal Mode Analysis (NMA):** Examines low-frequency vibrational modes of the protein.

- **Molecular Dynamics (MD):** Simulates the protein's motion over time, providing dynamic trajectories.

- **Conformational Ensemble:** Considers multiple protein conformations in docking calculations.

Application in Docking

Protein flexibility is particularly crucial in cases where induced fit or conformational changes occur upon ligand binding. It allows for a more realistic exploration of the ligand binding landscape, potentially revealing binding sites that rigid docking might overlook.

Mathematical Equations

The incorporation of protein flexibility can involve mathematical equations describing the motion of atoms in MD simulations or the vibrational modes in NMA. For example:

$$\text{MD Equation:} \quad m_i \frac{d^2 \mathbf{r}_i}{dt^2} = \mathbf{F}_i \tag{8.12}$$

This equation represents Newton's second law applied to an atom (i) in a molecular dynamics simulation.

Scoring Functions

Scoring functions for flexible docking should consider the dynamic nature of the protein, potentially incorporating terms related to conformational changes and flexibility.

Integrating protein flexibility enhances the predictive power of molecular docking simulations, providing a more accurate representation of ligand binding events.

8.2.8 Prediction of Binding Site Residues

Accurately predicting binding site residues is crucial for understanding molecular interactions and facilitating drug discovery efforts. Identifying the specific amino acids involved in ligand binding aids in the design of targeted therapeutics.

Computational Methods

Various computational methods are employed for predicting binding site residues:

- **Machine Learning Approaches:** Utilizing trained models on known binding site data to predict residues.

- **Sequence and Structure Conservation Analysis:** Identifying conserved regions across related proteins.

- **Ligand-Centric Approaches:** Analyzing ligand interactions to infer binding sites.

Mathematical Formulation

The prediction of binding site residues can be formulated as a classification problem, where each amino acid is classified as either part of the binding site or not. Mathematically, this can be expressed as:

$$\text{Prediction}(\mathbf{X}_i) = \begin{cases} \text{Binding Site,} & \text{if Score}(\mathbf{X}_i) \geq \text{Threshold} \\ \text{Non-Binding Site,} & \text{otherwise} \end{cases} \quad (8.13)$$

Here, $\mathbf{X}_i$ represents the features of the i-th amino acid, and $\text{Score}(\mathbf{X}_i)$ is a function that computes the likelihood of the amino acid being part of the binding site.

Integration with Structural Information

Predictions are often enhanced by incorporating structural information, such as solvent accessibility and spatial proximity to known ligands. This integration improves the accuracy of binding site residue prediction.

Validation and Benchmarking

Validation of binding site predictions involves comparing computational results with experimental data. Benchmarking against known structures helps assess the predictive performance of different methods.

Applications

Accurate binding site predictions find applications in virtual screening, rational drug design, and understanding protein function. They guide experimental efforts by focusing on key residues involved in ligand binding.

8.2.9　Water Molecules in Docking Studies

The role of water molecules in molecular docking studies is paramount, as they significantly influence ligand binding and protein conformation. Understanding the dynamic behavior of water molecules within the binding site is crucial for accurate predictions.

Dynamic Solvation in Binding Sites

Water molecules in the binding site contribute to solvation effects, influencing the energetics of ligand-protein interactions. Accounting for the dynamic nature of solvation is essential for accurate docking simulations.

Mathematical Representation

Incorporating water molecules in docking simulations involves considering their potential interactions with both the ligand and the protein. This can be mathematically represented as:

$$\text{Binding Energy} = \text{Ligand-Protein Interaction Energy} + \text{Solvation Energy}$$

$$(8.14)$$

Here, the solvation energy accounts for the contributions from water molecules, capturing their stabilizing or destabilizing effects.

Hydration Shell Analysis

Analyzing the hydration shell around the ligand-binding site provides insights into the preferred locations of water molecules. Molecular dynamics simulations and experimental techniques contribute to understanding the dynamic behavior of water in the binding pocket.

Water-Mediated Interactions

Water molecules can mediate interactions between the ligand and protein through hydrogen bonding or other polar interactions. Identifying these water-mediated interactions is crucial for a comprehensive understanding of binding mechanisms.

Challenges and Considerations

While the inclusion of water molecules enhances the realism of docking studies, it introduces computational challenges. Managing the dynamic nature of water molecules and accurately predicting their positions require advanced simulation techniques.

Applications

Consideration of water molecules in docking studies is particularly relevant in drug discovery, where accurate predictions of ligand binding and energetics are essential. Understanding the role of water enhances the rational design of therapeutically relevant compounds.

8.2.10   Quantitative Structure-Activity Relationship

Quantitative Structure-Activity Relationship (QSAR) is a powerful computational approach used in drug discovery to establish relationships between chemical structure and biological activity. QSAR models aid in predicting the biological activity of molecules based on their structural features.

Mathematical Foundation

The core of QSAR lies in mathematical models that correlate chemical descriptors with biological activities. The general form of a QSAR equation is:

$$\text{Activity} = f(\text{Descriptors}) \tag{8.15}$$

Here, "Activity" represents the biological activity of a molecule, and "Descriptors" include various molecular properties such as size, shape, electronegativity, and other physicochemical characteristics.

Chemical Descriptors

Chemical descriptors are quantitative representations of molecular features. They can be categorized into molecular, topological, electronic, and spatial descriptors. Examples include molecular weight, logP (partition coefficient), and the number of hydrogen bond donors/acceptors.

Model Development

Developing a QSAR model involves selecting relevant descriptors, obtaining experimental activity data, and using statistical methods to establish a correlation. Regression analysis is commonly employed to derive the coefficients of the model.

Validation

Validation is a critical step in ensuring the reliability of QSAR models. Techniques like cross-validation and external validation using independent datasets assess the model's predictive performance.

Applications

QSAR finds applications in virtual screening, lead optimization, and understanding structure-activity relationships. It aids medicinal chemists in prioritizing compounds for synthesis, thereby saving time and resources.

Limitations

Despite its utility, QSAR has limitations, including the need for high-quality experimental data, the complexity of biological systems, and potential overfitting. Careful consideration and validation are essential for robust QSAR models.

Emerging Trends

Advancements in machine learning techniques, availability of larger datasets, and integration with other computational methods are contributing to the evolution of QSAR, enhancing its predictive power and applicability.

8.2.11   Understanding Allosteric Sites

Allosteric sites on proteins play a crucial role in modulating their activity and function. Unlike active sites where substrates bind directly, allosteric sites are distinct regions that, when bound by molecules (allosteric modulators), induce conformational changes affecting the protein's activity.

Definition

An allosteric site is a site on a protein away from the active site where regulatory molecules, often small ligands, can bind and influence the protein's function. This interaction can lead to either positive allosteric modulation (increased activity) or negative allosteric modulation (decreased activity).

Mathematical Representation

Mathematically, allosteric modulation can be represented using equations that describe the interaction between the protein, the substrate, and the allosteric modulator. For example, in a simplified form:

$$\text{Protein (R)} + \text{Substrate (S)} \xtofrom[\text{Allosteric Modulator (M)}]{\text{Active site}} \text{Protein-Substrate Complex (RS)}$$

$$(8.16)$$

This equation illustrates the dynamic equilibrium between the different states of the protein.

Conformational Changes

Upon allosteric modulation, conformational changes occur in the protein structure. This structural shift can impact the active site's accessibility or alter the protein's affinity for its substrate.

Biological Significance

Understanding allosteric sites is crucial for drug discovery and design. Allosteric modulators can be targeted to enhance or inhibit specific protein functions, providing a more nuanced and selective approach compared to targeting active sites directly.

Examples

Notable examples of allosteric modulation include the binding of 2,3-BPG to hemoglobin, affecting its oxygen-binding affinity, and the binding of allosteric inhibitors to enzymes, regulating their activity.

Experimental Techniques

Experimental techniques such as X-ray crystallography, NMR spectroscopy, and computational simulations are employed to identify and study allosteric sites, providing insights into their mechanisms.

Therapeutic Implications

Targeting allosteric sites has therapeutic implications, as drugs designed to modulate protein function through allosteric regulation can offer specificity and reduced side effects compared to traditional orthosteric drugs.

Challenges and Future Directions

Despite their significance, allosteric sites present challenges in terms of prediction and design. Ongoing research aims to deepen our understanding of allosteric mechanisms and develop innovative strategies for therapeutic intervention.

8.2.12 Predicting Ligand-Induced Conformational Changes

The prediction of ligand-induced conformational changes in proteins is a critical aspect of molecular docking studies. Understanding how ligand binding influences the structure of a protein is essential for uncovering the molecular mechanisms underlying biological processes and for drug discovery applications.

Significance

Ligand-induced conformational changes refer to alterations in the three-dimensional structure of a protein induced by the binding of a ligand. These changes can impact the protein's function, stability, and interaction with other molecules. Predicting these changes is crucial for elucidating the dynamic nature of protein-ligand interactions.

Mathematical Models

Mathematical models are employed to predict ligand-induced conformational changes. One common approach involves molecular dynamics simulations, where the equations of motion for atoms in the protein and ligand are numerically solved to simulate their behavior over time. The resulting trajectories provide insights into dynamic changes in protein structure.

Energy Landscape

The prediction process often involves analyzing the energy landscape of the protein-ligand complex. This landscape represents the potential energy of the system as a function of the protein's conformational states. Minima in the landscape correspond to stable conformations, and barriers indicate energy barriers that must be overcome for conformational changes.

Experimental Validation

Experimental techniques such as X-ray crystallography, NMR spectroscopy, and cryo-electron microscopy are used to validate predicted conformational changes. These techniques provide high-resolution structural information, confirming the accuracy of computational predictions.

Role in Drug Discovery

In drug discovery, predicting ligand-induced conformational changes is crucial for designing molecules that can selectively modulate a target protein's activity. Understanding how ligands influence protein structure aids in the rational design of drugs with improved binding affinity and specificity.

Challenges

Predicting ligand-induced conformational changes faces challenges, including the need for computational resources for extensive simulations and the inherent complexity of protein dynamics. Overcoming these challenges requires advancements in simulation algorithms and increased understanding of the underlying biological processes.

Future Directions

Future research directions involve the development of more efficient computational algorithms, incorporation of machine learning approaches, and collaborative efforts between experimental and computational biologists to refine and validate predictive models.

Conclusion

Predicting ligand-induced conformational changes is a dynamic and evolving field with profound implications for understanding biological processes and advancing drug discovery. Continued interdisciplinary efforts are essential for pushing the boundaries of our predictive capabilities.

8.2.13 Examples of Structural Bioinformatics-Driven Docking Studies

Structural bioinformatics plays a pivotal role in guiding molecular docking studies, providing valuable insights into the structural features of biological macromolecules. Here are examples illustrating the application of structural bioinformatics in various docking studies:

Protein-Protein Interaction:

Objective: Investigating the binding mechanism between two protein partners involved in signal transduction.

Approach: Structural bioinformatics analysis of the interacting proteins revealed key residues at the interface. Docking simulations were performed considering the identified hotspots, leading to the prediction of the most energetically favorable binding mode.

Ligand Binding Site Prediction:

Objective: Identifying potential binding sites on a protein for a novel small-molecule ligand.

Approach: Structural bioinformatics tools were employed to analyze the protein's surface features and predict plausible binding sites. Docking simulations with diverse ligands validated the predicted binding sites, aiding in ligand design.

Drug Repurposing:

Objective: Repurposing an existing drug for a different target by understanding its interaction with various proteins.

Approach: Structural bioinformatics analysis of known drug-target complexes provided insights into common binding motifs. Docking studies were performed with the drug against multiple targets, revealing potential off-target interactions.

Antigen-Antibody Docking:

Objective: Predicting the binding mode between an antigenic peptide and an antibody for vaccine design.

Approach: Structural bioinformatics analysis of the antigenic peptide and antibody variable regions guided the docking simulations. The study resulted in the identification of the most probable binding orientation critical for eliciting an immune response.

Enzyme Substrate Specificity:

Objective: Understanding the substrate specificity of an enzyme and predicting interactions with different substrates.

Approach: Structural bioinformatics analysis of the enzyme's active site and substrate-binding pocket guided the selection of substrates for docking studies. The results provided insights into the enzyme's preferences for specific substrate structures.

Membrane Protein Interactions:

Objective: Investigating the interaction between membrane proteins to understand cell signaling pathways.

Approach: Structural bioinformatics tools were employed to model the transmembrane regions and predict interaction interfaces. Docking simulations with membrane-embedded proteins helped elucidate the nature of their interactions.

8.2.14 Challenges and Future Directions

Despite the advancements, challenges persist in structural bioinformatics-driven docking studies. Accuracy in predicting protein flexibility, handling membrane proteins, and addressing the influence of solvents are ongoing challenges. Future directions involve the integration of artificial intelligence and machine learning approaches to enhance the predictive capabilities of structural bioinformatics tools in molecular docking.

Chapter 9

Challenges and Future Directions

9.1 Current Challenges

Molecular docking, while a powerful tool in drug discovery, faces several challenges that impede its efficiency and accuracy. Addressing these challenges is crucial for advancing the field and unlocking its full potential. This section explores the current challenges in molecular docking and discusses potential strategies to overcome them.

9.1.1 Protein Flexibility and Conformational Changes

Protein flexibility and conformational changes play a significant role in molecular docking simulations. Understanding and modeling the dynamic nature of proteins are crucial for accurately predicting ligand binding.

Mathematical Formula for Protein Flexibility

A mathematical formula representing protein flexibility might involve considering different protein conformations:

$$\text{Protein\_Flexibility} = \arg\min_{\text{conformations}} \left(\sum_{i=1}^{N} \text{Energy}(Protein_{\text{original}} \\ + \text{Conformation}_i + \text{Ligand}) \right) \tag{9.1}$$

Here, Energy represents the scoring function incorporating protein conformations.

Chemical Formula for Conformational Changes

The representation of protein flexibility in a chemical formula:

$$\text{Protein}_{\text{original}} + \text{Conformation}_i + \text{Ligand} \longrightarrow \text{Protein}_{\text{flexible}} \tag{9.2}$$

This indicates the transformation of the original protein with a specific conformation upon ligand binding.

Accurate modeling of protein flexibility is essential for capturing the nuances of ligand-protein interactions in molecular docking studies.

9.1.2   Treatment of Solvent Effects

The inclusion of solvent effects is crucial in molecular docking simulations to account for the environment's influence on ligand-receptor interactions. Various methods are employed to address solvent effects during the docking process.

Mathematical Formula for Solvent Effects

A mathematical formula representing solvent effects might involve considering solvation energy:

$$\text{Total\_Energy} = \text{Binding\_Energy} + \text{Solvation\_Energy} \tag{9.3}$$

Here, Binding_Energy represents the energy associated with ligand-receptor binding, and Solvation_Energy accounts for the interaction between the solute and solvent.

Chemical Formula for Solvent Effects

The representation of solvent effects in a chemical formula:

$$\text{Ligand} + \text{Receptor}_{\text{in\_solvent}} \longrightarrow \text{Ligand}-\text{Receptor\_Complex} + \text{Solvent} \quad (9.4)$$

This equation illustrates the formation of the ligand-receptor complex within the solvent environment.

Accurate treatment of solvent effects enhances the realism of molecular docking simulations, providing insights into ligand binding in biologically relevant conditions.

9.1.3   Handling Membrane Proteins

Molecular docking involving membrane proteins introduces additional challenges due to the presence of lipid bilayers. Proper handling of membrane proteins is essential for realistic simulations.

Mathematical Considerations

Incorporating membrane effects in the scoring function may involve terms related to the lipid environment:

$$\text{Total\_Energy} = \text{Binding\_Energy} + \text{Membrane\_Energy} \quad (9.5)$$

Here, Binding_Energy represents the ligand-receptor binding energy, and Membrane_Energy accounts for interactions within the lipid bilayer.

Chemical Representation

The interaction of a ligand with a membrane protein can be chemically represented as:

$$\text{Ligand} + \text{Membrane\_Protein} \longrightarrow \text{Ligand}-\text{Membrane\_Protein\_Complex} \quad (9.6)$$

This equation signifies the formation of a complex between the ligand and the membrane protein within the lipid bilayer.

Effectively addressing membrane proteins in docking simulations contributes to a more accurate representation of biological processes occurring in cell membranes.

9.1.4 Scoring Function Accuracy

The accuracy of a scoring function in molecular docking is crucial for predicting reliable ligand-binding affinities. The scoring function's performance can be assessed using various metrics, and one common metric is the Root Mean Square Deviation (RMSD) between predicted and experimental binding affinities:

$$\text{RMSD} = \sqrt{\frac{\sum_{i=1}^{N}(\text{Predicted Affinity}_i - \text{Experimental Affinity}_i)^2}{N}} \quad (9.7)$$

Here, Predicted Affinity$_i$ represents the predicted binding affinity for the i-th ligand, and Experimental Affinity$_i$ represents the corresponding experimental binding affinity. The summation is performed over all ligands (N) in the dataset.

A low RMSD value indicates a close agreement between predicted and experimental affinities, reflecting the accuracy of the scoring function.

Scoring function accuracy is paramount for the success of molecular docking simulations, especially in the context of drug discovery.

9.1.5 Treatment of Metal Ions and Cofactors

In molecular docking simulations, the treatment of metal ions and cofactors is essential for accurate predictions of ligand binding. The presence of metal ions can significantly influence the binding affinity and coordination geometry.

One approach to model metal-ligand interactions is by using a force field that explicitly considers metal parameters. The interaction energy ($E_{\text{metal-ligand}}$) can be represented as:

$$E_{\text{metal-ligand}} = \frac{1}{2}k(d - d_0)^2 \quad (9.8)$$

Here, k is the force constant, d is the metal-ligand distance, and d_0 is the equilibrium distance.

For cofactors, their flexible nature may require special handling, and the choice of parameters depends on the specific characteristics of the cofactor. Properly representing metal ions and cofactors in the docking process enhances the accuracy of binding predictions, especially in cases where these elements play a crucial role in ligand recognition.

9.1.6 Incorporating Allosteric Interactions

Allosteric interactions play a crucial role in ligand binding and can modulate the activity of proteins. In molecular docking, incorporating allosteric effects enhances the accuracy of predictions, considering the influence of distal binding sites.

A common approach to model allosteric interactions is to introduce an allosteric modulation term in the scoring function. The modified interaction energy ($E_{\text{allosteric-ligand}}$) can be represented as:

$$E_{\text{allosteric-ligand}} = \alpha \cdot E_{\text{binding}} + \beta \cdot E_{\text{allosteric}} \qquad (9.9)$$

Here, E_{binding} is the binding energy without allosteric effects, $E_{\text{allosteric}}$ is the allosteric modulation term, and α and β are weighting factors.

The allosteric modulation term can be further defined based on specific properties or interactions involved in the allosteric regulation. This inclusion allows for a more comprehensive representation of ligand-protein interactions, considering both primary binding sites and allosteric regulation.

9.1.7 Handling Large-Scale Virtual Screening

Large-scale virtual screening involves screening a vast number of compounds to identify potential ligands efficiently. Efficient handling of such datasets requires optimized algorithms and parallel processing.

One common approach is to parallelize the virtual screening process using distributed computing. The overall score (S_{total}) for each ligand can be calculated as:

$$S_{\text{total}} = \sum_{i=1}^{N} S_i \tag{9.10}$$

Here, N is the total number of ligands, and S_i represents the score for each ligand.

To parallelize the scoring process, the total score can be distributed across multiple processors or computing nodes:

$$S_{\text{total}} = \sum_{j=1}^{M} \sum_{i=1}^{N/M} S_{(j-1)\cdot(N/M)+i} \tag{9.11}$$

Where M is the number of processors or nodes used for parallelization.

Efficient parallelization enables the rapid screening of large compound libraries, making it feasible to explore extensive chemical spaces in drug discovery efforts.

9.1.8   Accounting for Protein-Ligand Binding Pathways

Understanding the binding pathways of protein-ligand interactions is crucial for predicting the binding affinity accurately. The binding pathway can be represented using the potential of mean force (PMF) along a reaction coordinate.

$$PMF = -k_B T \ln P \tag{9.12}$$

Here, k_B is the Boltzmann constant, T is the temperature, and P is the probability distribution of the ligand along the reaction coordinate.

To account for different binding pathways, molecular dynamics simulations can be employed to sample various conformations and interactions. The free energy landscape (G) can be determined from the PMF:

$$G = -k_B T \ln \int P \, dX \tag{9.13}$$

Where X represents the reaction coordinate.

Analyzing the free energy landscape provides insights into the energetically favorable binding pathways and aids in the accurate prediction of binding affinities.

9.1.9 Improving Predictions for Weak Binders

Accurately predicting the binding affinity for weak binders presents a challenge in molecular docking. To enhance predictions, adjustments can be made to the scoring functions, particularly for interactions involving weak binding.

Modified Scoring Function

One approach is to introduce adjustments to the scoring function to better capture the subtle interactions in weak binding scenarios. A modified scoring function can be expressed as:

$$\text{Score} = w_1 \cdot \text{VDW Energy} + w_2 \cdot \text{Electrostatic Energy} + w_3 \qquad (9.14)$$
$$\cdot \text{Hydrogen Bonding Energy} + \ldots + w_n \cdot \text{Additional Terms}$$

Here, $w_1, w_2, \ldots, w_n$ are weights assigned to different energy terms. These weights can be adjusted to prioritize certain interactions, giving more emphasis to those contributing significantly to weak binding.

Chemical Formula

In chemical formulas, the modified scoring function can be represented as:

$$\text{Score} = w_1 \cdot \text{VDW Energy} + w_2 \cdot \text{Electrostatic Energy} + w_3 \qquad (9.15)$$
$$\cdot \text{Hydrogen Bonding Energy} + \ldots + w_n \cdot \text{Additional Terms}$$

Adjusting the weights allows for fine-tuning the scoring function to improve predictions for weak binders.

9.1.10   Handling Conformational Sampling

Accurate molecular docking requires effective sampling of the conformational space to explore various ligand-receptor configurations. Efficient conformational sampling is crucial for capturing potential binding modes.

Monte Carlo Method

One widely used approach for conformational sampling is the Monte Carlo method. It involves generating random ligand conformations and evaluating their interactions with the receptor. The probability of accepting a new conformation is determined by the Metropolis criterion.

$$P(\text{Accept}) = \min\left(1, \exp\left(-\frac{\Delta E}{kT}\right)\right) \tag{9.16}$$

Here, ΔE is the energy difference between the new and old conformations, k is the Boltzmann constant, and T is the temperature.

Chemical Formula

In chemical formulas, the Metropolis criterion can be expressed as:

$$P(\text{Accept}) = \min\left(1, \exp\left(-\frac{\Delta E}{kT}\right)\right) \tag{9.17}$$

Utilizing Monte Carlo methods enhances the exploration of ligand conformational space during docking simulations.

9.1.11   Incorporating Quantum Mechanical Effects

To enhance the accuracy of molecular docking simulations, incorporating quantum mechanical (QM) effects becomes essential, especially when dealing with systems where electronic interactions play a significant role.

QM/MM Approaches

Quantum mechanics/molecular mechanics (QM/MM) methods are commonly employed to merge the accuracy of quantum mechanical calculations with the efficiency of molecular mechanics simulations. The total energy (E_{total}) in a QM/MM approach is expressed as:

$$E_{\text{total}} = E_{\text{QM}} + E_{\text{MM}} + E_{\text{QM/MM}} \tag{9.18}$$

Here, E_{QM} represents the quantum mechanical energy, E_{MM} is the molecular mechanics energy, and $E_{\text{QM/MM}}$ accounts for the interactions between the QM and MM regions.

Chemical Formula

In chemical formulas, the QM/MM total energy can be represented as:

$$E_{\text{total}} = E_{\text{QM}} + E_{\text{MM}} + E_{\text{QM/MM}} \tag{9.19}$$

Incorporating QM effects enables a more accurate description of electronic properties during ligand-receptor interactions.

9.1.12 Addressing Ligand Flexibility

Consideration of ligand flexibility is crucial for accurate molecular docking simulations, as it allows the ligand to adopt different conformations during binding interactions.

Flexible Ligand Representation

A common approach to address ligand flexibility is to use multiple ligand conformations or an ensemble of structures. The flexible ligand representation can be denoted as:

$$\text{Ligand} \equiv \{\text{Conformation}_1, \text{Conformation}_2, \ldots, \text{Conformation}_n\} \tag{9.20}$$

Here, $\{\}$ represents a set, and each Conformation$_i$ is a different possible conformation of the ligand.

Mathematical Formula

In mathematical terms, the ligand flexibility can be incorporated into the docking score as a weighted sum of different conformations:

$$\text{Docking Score} = \sum_{i=1}^{n} w_i \cdot \text{Score}(\text{Receptor}, \text{Conformation}_i) \qquad (9.21)$$

Here, w_i represents the weight associated with each ligand conformation, and $\text{Score}(\text{Receptor}, \text{Conformation}_i)$ is the docking score for a specific ligand conformation.

Chemical Formula

In chemical formulas, ligand flexibility can be represented as:

$$\text{Ligand} \equiv \{\text{Conformation}_1, \text{Conformation}_2, \ldots, \text{Conformation}_n\} \qquad (9.22)$$

Addressing ligand flexibility is essential for capturing the dynamic nature of ligand-receptor interactions in molecular docking simulations.

9.1.13   Data Standardization and Reproducibility

Ensuring data standardization is crucial for reproducibility in molecular docking studies. Standardization helps in maintaining consistency across datasets, facilitating comparison, and ensuring the reliability of results.

Standardization Process

The process of data standardization involves normalizing various parameters, such as ligand and receptor structures, affinity values, and experimental conditions. This can be represented as:

$$\text{Standardized Data} = \text{Normalize(Raw Data)} \tag{9.23}$$

The Normalize function adjusts the data to a common scale or format, minimizing variability introduced by experimental differences.

Mathematical Formula

Mathematically, data standardization can be expressed as:

$$\text{Standardized Value}_i = \frac{\text{Raw Value}_i - \text{Raw } \bar{\text{Values}}}{\sigma_{\text{Raw Values}}} \tag{9.24}$$

Here, Raw $\bar{\text{Values}}$ is the mean, and $\sigma_{\text{Raw Values}}$ is the standard deviation of the raw values.

Chemical Formula

In chemical terms, data standardization can be represented as:

$$\text{Standardized Data} = \text{Normalize(Raw Data)} \tag{9.25}$$

Standardizing data ensures that molecular docking results are comparable and reproducible, contributing to the robustness of computational studies.

9.1.14 Enhancing User Accessibility

In the development of molecular docking tools, prioritizing user accessibility is essential to ensure a wide range of researchers can benefit from these computational techniques.

User Interface Improvements

Enhancements to the user interface (UI) play a crucial role in making molecular docking tools more accessible. This involves creating intuitive menus, visualizations, and interactive features to simplify the user experience.

Mathematical Formula

To quantify user accessibility improvements, a mathematical formula may be defined based on user feedback and efficiency metrics:

$$\text{Accessibility Score} = \frac{\text{Number of Positive User Feedback}}{\text{Total User Interactions}} \times 100 \quad (9.26)$$

Here, the Accessibility Score is a percentage calculated from the ratio of positive user feedback to the total number of user interactions, providing a quantitative measure of user satisfaction.

Chemical Formula

In the context of chemical informatics, the enhancement of user accessibility can be expressed as:

$$\text{Improved Accessibility} = \text{User-friendly UI} + \text{Intuitive Features} \quad (9.27)$$

Improving the accessibility of molecular docking tools contributes to the democratization of computational methods, allowing researchers with varying expertise to leverage these powerful techniques.

9.1.15   Integrating Experimental Data with Computational Models

The integration of experimental data into computational models is a pivotal aspect of refining the accuracy and reliability of molecular docking predictions.

Data Fusion

To combine experimental data with computational results, a data fusion approach can be employed. This involves merging information from various sources to create a comprehensive dataset for model training and validation.

Mathematical Formula

A mathematical formula for integrating experimental and computational data can be expressed as follows:

$$\text{Integrated Data} = f(\text{Experimental Data}, \text{Computational Predictions}) \quad (9.28)$$

Here, f represents a function that combines experimental data and computational predictions to generate a unified dataset.

Chemical Formula

In the context of molecular structures, the integration can be symbolized as:

$$\text{Integrated Structure} = \text{Experimental Structure} + \text{Computational Predicted Structure}$$
$$(9.29)$$

This symbolizes the amalgamation of experimental and computational structural information to enhance the overall understanding of molecular interactions.

Integrating experimental data with computational models bridges the gap between theoretical predictions and real-world observations, contributing to the refinement of molecular docking methodologies.

9.1.16 Addressing Bias in Training Datasets

Addressing bias in training datasets is crucial for ensuring the generalizability and fairness of machine learning models in molecular docking.

Bias Correction

To correct bias in training datasets, various techniques can be employed. One common approach is to reweight the samples based on their underrepresented or overrepresented nature. This can be mathematically expressed as:

$$\text{Reweighted Sample} = \text{Original Sample} \times \frac{\text{Desired Distribution}}{\text{Actual Distribution}} \quad (9.30)$$

Here, the desired distribution represents the unbiased distribution, and the actual distribution is the distribution present in the original dataset.

Chemical Formula

In the realm of chemical structures, addressing bias might involve adjusting the representation of certain molecular features. This can be symbolized as:

$$\text{Adjusted Structure} = \text{Original Structure} + \text{Bias Correction} \qquad (9.31)$$

This equation signifies the modification of the original structure to account for biases identified in the training dataset.

Addressing bias in training datasets is essential for building machine learning models that can provide accurate predictions across diverse molecular scenarios.

9.1.17   Improving Predictive Capabilities for Challenging Targets

Improving the predictive capabilities of molecular docking models for challenging targets involves addressing specific issues associated with these targets.

Enhanced Scoring Functions

One approach to enhance predictive capabilities is through the development of advanced scoring functions. A potential enhancement can be achieved by incorporating additional terms that capture the unique features of challenging targets:

$$\text{Enhanced Score} = \text{Base Score} + \alpha \cdot \text{Additional Term} \qquad (9.32)$$

Here, α represents a weight parameter controlling the contribution of the additional term.

Chemical Formula

In a chemical context, the improvement in predictive capabilities may involve adjusting the representation of certain ligand or receptor features. This adjustment can be symbolized as:

$$\text{Improved Molecule} = \text{Original Molecule} + \text{Enhancement} \tag{9.33}$$

The equation illustrates the modification of the original molecule to improve its compatibility with challenging targets.

Improving predictive capabilities for challenging targets is an ongoing research area, aiming to broaden the applicability of molecular docking across diverse biological systems.

9.1.18 Incorporating Thermodynamic Considerations

The incorporation of thermodynamic considerations in molecular docking models is essential for a more accurate representation of ligand-receptor interactions.

Gibbs Free Energy Calculation

One way to incorporate thermodynamics is by calculating the Gibbs free energy (ΔG) associated with the binding process:

$$\Delta G = \Delta H - T\Delta S \tag{9.34}$$

Here, ΔH is the enthalpy change, ΔS is the entropy change, and T is the absolute temperature.

Chemical Formula

In a chemical context, the thermodynamic considerations may be reflected in the representation of chemical reactions. For example, the Gibbs free energy change (ΔG) for a reaction can be expressed as:

$$\text{A} + \text{B} \rightleftharpoons \text{C} \quad \Delta G = \Delta G^\circ + RT \ln\left(\frac{[\text{C}]}{[\text{A}][\text{B}]}\right) \tag{9.35}$$

Where ΔG° is the standard Gibbs free energy change, R is the ideal gas constant, T is the absolute temperature, and [A], [B], and [C] are the concentrations of the respective species.

Incorporating thermodynamic considerations enhances the predictive power of molecular docking models by accounting for the energetics of ligand-receptor interactions.

9.1.19   Navigating Interactions in Dynamic Cellular Environments

Understanding molecular interactions in dynamic cellular environments is crucial for predicting ligand behavior within living systems.

Diffusion and Cellular Environment

The movement of ligands within a cell can be described by diffusion equations. One such equation is Fick's second law:

$$\frac{\partial C}{\partial t} = D\nabla^2 C \tag{9.36}$$

Here, C is the concentration of the ligand, t is time, D is the diffusion coefficient, and ∇^2 is the Laplacian operator.

Chemical Formula

In a chemical context, the dynamic cellular environment may be represented in a chemical reaction involving ligands and cellular components:

$$\text{Ligand} + \text{Cellular Component} \rightleftharpoons \text{Ligand-Cellular Complex} \tag{9.37}$$

This reaction indicates the dynamic nature of ligand interactions within the cellular environment.

Navigating interactions in dynamic cellular environments requires considering factors such as diffusion, cellular structures, and biochemical transformations. Incorporating these aspects into molecular docking models improves their relevance to real-world biological systems.

9.1.20 Balancing Speed and Accuracy in Docking Calculations

Achieving an optimal balance between computational speed and docking accuracy is a critical consideration in molecular docking simulations.

Trade-off Formula

One way to express the trade-off between speed and accuracy is through a mathematical formula. Let S represent computational speed, and A represent accuracy. A common trade-off formula might be:

$$\text{Trade-off} = k \cdot \frac{1}{S} + (1 - k) \cdot A \tag{9.38}$$

Here, k is a parameter controlling the balance between speed and accuracy. Adjusting k allows researchers to prioritize one aspect over the other.

Chemical Formula

In a chemical context, the balance between speed and accuracy may be likened to finding the optimal conditions for a reaction:

$$\text{OptimalConditions} \xrightarrow{\text{Speed}} \text{Products}_{\text{Fast}} + [\text{Accuracy}]\,\text{Products}_{\text{Accurate}} \tag{9.39}$$

This chemical formula metaphorically captures the idea that different conditions may favor speed or accuracy.

Achieving an effective balance involves fine-tuning parameters and algorithms to suit the specific goals of the docking study. Striking the right balance ensures that computational resources are used efficiently without compromising the reliability of the results.

9.1.21   Promoting Open Science and Collaboration

Open science and collaboration play pivotal roles in advancing molecular docking research. Embracing transparency, sharing methodologies, and fostering collaboration contribute to the collective progress of the scientific community.

Mathematical Formula for Collaboration Index

Quantifying collaboration impact can be expressed with a mathematical formula, where C represents the collaboration index:

$$C = \frac{\text{Number of Collaborators}}{\text{Total Publications}} \tag{9.40}$$

This formula provides a metric for the level of collaborative engagement in research projects. A higher collaboration index signifies a more interconnected and collaborative scientific community.

Chemical Formula of Collaboration

In a chemical metaphor, collaboration can be likened to the synergistic interaction of molecular entities:

$$\text{Researcher}_1 + \text{Researcher}_2 \xrightarrow{\text{Collaboration}} \text{Scientific}_\text{A}\text{dvancement} \tag{9.41}$$

This chemical formula symbolizes the combined efforts of researchers leading to scientific advancement.

Promoting open science involves sharing data, codes, and methodologies openly, allowing others to reproduce and validate findings. Collaborative initiatives enhance the reliability and generalizability of molecular docking research.

Encouraging a culture of openness and collaboration is essential for the continued growth and innovation in the field.

9.2 Emerging Trends

The field of molecular docking is dynamic, with continuous advancements shaping its landscape. This section explores emerging trends that are influencing the future directions of molecular docking, providing a glimpse into the evolving methodologies and technologies that researchers are embracing.

9.2.1 Deep Learning in Scoring Functions

Deep learning has emerged as a powerful tool in enhancing scoring functions for molecular docking. Leveraging neural networks, especially deep architectures, contributes to more accurate predictions of binding affinities.

Neural Network Scoring Function

A neural network-based scoring function can be represented as follows:

$$\text{Score}(\text{Ligand}, \text{Receptor}) = \sigma \left(\sum_{i=1}^{N} w_i \cdot f_i(\text{Ligand}, \text{Receptor}) + b \right) \quad (9.42)$$

Here:

N is the number of features considered,

w_i represents the weights assigned to each feature,

$f_i(\text{Ligand}, \text{Receptor})$ denotes the feature extracted from the ligand-receptor complex,

b is the bias term,

σ is the activation function (e.g., sigmoid or ReLU).

The neural network learns optimal weights during the training process, adapting to complex patterns and relationships in the data.

Chemical Metaphor for Deep Learning

In a chemical metaphor, deep learning can be compared to the intricate reactions guiding molecular transformations:

$$\text{Ligand} + \text{Receptor} \xrightarrow{\text{Deep Learning}} \text{High Precision Binding Affinity} \qquad (9.43)$$

This chemical formula symbolizes the transformative impact of deep learning in refining the accuracy of binding affinity predictions.

Deep learning-based scoring functions demonstrate promising results in capturing intricate interactions within ligand-receptor complexes, paving the way for more reliable molecular docking simulations.

9.2.2   Hybrid Methods Integrating Experimental Data

Hybrid methods that seamlessly integrate computational predictions with experimental data have become increasingly valuable in enhancing the accuracy and reliability of molecular docking studies.

Combined Scoring Function

One common approach involves combining the computational scoring function (S_{comp}) with experimentally derived terms (S_{exp}) to form a hybrid scoring function (S_{hybrid}):

$$S_{\text{hybrid}} = w_{\text{comp}} \cdot S_{\text{comp}} + w_{\text{exp}} \cdot S_{\text{exp}} \qquad (9.44)$$

Here:

w_{comp} is the weight assigned to the computational term,

w_{exp} is the weight assigned to the experimental term.

This formula allows for the seamless integration of computational predictions and experimental measurements, providing a more comprehensive view of the ligand-receptor interactions.

Chemical Metaphor for Hybrid Integration

In a chemical metaphor, the hybrid integration of computational and experimental data can be likened to a synergistic reaction:

Computational Prediction (9.45)

$+$ Experimental Data $\xrightarrow{\text{Hybrid Integration}}$ Enhanced Binding Affinity

This metaphor reflects the collaborative nature of hybrid methods, where the strengths of both computational and experimental approaches synergize to improve the accuracy of binding affinity predictions.

Hybrid methods integrating computational and experimental data offer a robust framework for refining molecular docking simulations, addressing the limitations of individual methods.

9.2.3 Advancements in Quantum Mechanical Docking

Recent advancements in molecular docking have seen the integration of quantum mechanics principles, providing a more accurate representation of the electronic structure during ligand-receptor interactions.

Quantum Mechanical Scoring Function

The key innovation lies in the incorporation of quantum mechanical effects into the scoring function. The quantum mechanical scoring function (S_{QM}) can be expressed as:

$$S_{\text{QM}} = \langle \Psi_{\text{complex}} | \hat{H} | \Psi_{\text{complex}} \rangle - \langle \Psi_{\text{ligand}} | \hat{H} | \Psi_{\text{ligand}} \rangle - \langle \Psi_{\text{receptor}} | \hat{H} | \Psi_{\text{receptor}} \rangle$$
(9.46)

Here:

$\langle \Psi_{\text{complex}} |, \langle \Psi_{\text{ligand}} |, \langle \Psi_{\text{receptor}} |$

represent the wavefunctions of the complex, ligand, and receptor,

$\hat{H}$ is the quantum mechanical Hamiltonian operator.

This scoring function captures the quantum mechanical contributions to the binding energy, providing a more accurate assessment of the ligand-receptor interaction.

Chemical Representation of Quantum Docking

In a chemical representation, quantum mechanical docking can be metaphorically illustrated as an electronic dance between the ligand and receptor:

$$\text{Ligand} \underset{}{\overset{\text{Quantum Mechanical Dance}}{\rightleftharpoons}} \text{Receptor} \qquad (9.47)$$

This metaphor emphasizes the dynamic and quantum nature of the interaction, highlighting the electronic intricacies involved.

Advancements in quantum mechanical docking represent a significant step towards achieving higher accuracy in predicting binding affinities, especially for systems where electronic effects play a crucial role.

9.2.4   Expanding Applications in Fragment-Based Drug Design

Fragment-based drug design (FBDD) has evolved as a powerful strategy for lead discovery and optimization. Recent advancements have expanded the applications of FBDD, combining experimental and computational approaches.

Fragment-Based Binding Affinity Prediction

The binding affinity (BA) of a fragment (F) to a target (T) can be predicted using a scoring function:

$$BA_{\text{predicted}} = \sum_i^n \left(\text{Score}_{\text{fragment}}(F_i, T) + \text{Score}_{\text{context}}(F_i, T) \right) \qquad (9.48)$$

Here, $\text{Score}_{\text{fragment}}$ evaluates the interaction of the individual fragment with the target, and $\text{Score}_{\text{context}}$ considers the context of neighboring fragments.

Chemical Representation of Fragment-Based Drug Design

In a chemical representation, the process of fragment-based drug design can be illustrated as a puzzle assembly:

$$\text{Fragments} \xrightarrow{\text{FBDD}} \text{Lead Compounds} \tag{9.49}$$

This metaphor emphasizes the systematic construction of lead compounds from smaller, chemically relevant fragments.

Integration of Machine Learning

To enhance fragment-based drug design, machine learning algorithms are employed to predict binding affinities and guide the selection of promising fragments. The machine learning model can be represented as:

$$BA_{\text{predicted}} = \text{ML\_Model}(F, T) \tag{9.50}$$

This integration facilitates efficient exploration of chemical space and accelerates lead optimization.

Expanding applications in fragment-based drug design showcase its versatility in identifying novel, high-quality leads, making it a valuable approach in modern drug discovery.

9.2.5 Incorporating Machine Learning in Ligand Design

The integration of machine learning (ML) techniques in ligand design has revolutionized the drug discovery process. ML models are employed to predict ligand properties, enhance molecular design, and expedite the identification of potent candidates.

Machine Learning Model for Ligand Property Prediction

Consider a machine learning model (ML_Model) that predicts a ligand property (Property) based on its molecular features:

$$\text{Property}_{\text{predicted}} = \text{ML\_Model}(\text{Molecular Features}) \tag{9.51}$$

Here, Molecular Features include descriptors such as molecular weight, lipophilicity, and electronic properties.

Chemoinformatics-Based Ligand Design

Chemoinformatics techniques leverage ML models to guide ligand design. The ligand design process can be expressed as:

$$\text{Lead Compound} \xrightarrow{\text{Chemoinformatics}} \text{Ligand Design} \xrightarrow{\text{ML\_Model}} \text{Optimized Ligand} \tag{9.52}$$

Chemoinformatics assists in exploring chemical space efficiently, identifying ligand modifications, and predicting their impact on properties.

Generative Models for Ligand Generation

Generative models, such as variational autoencoders (VAEs) or generative adversarial networks (GANs), can be employed for de novo ligand design. The generative process is represented as:

$$\text{Noise} \xrightarrow{\text{Generative Model}} \text{Novel Ligand Structures} \tag{9.53}$$

Generative models facilitate the exploration of diverse ligand candidates beyond existing chemical knowledge.

Incorporating machine learning in ligand design enhances the precision and efficiency of the drug discovery pipeline, driving the discovery of novel therapeutic agents.

9.2.6   GPU Acceleration for High-Performance Docking

To address the computational demands of molecular docking simulations, the integration of Graphics Processing Units (GPUs) has emerged as a powerful solution. GPUs provide parallel processing capabilities, significantly accelerating docking calculations.

Parallelization of Docking Algorithms

Consider a docking algorithm (Docking_Algorithm) that can be parallelized for GPU acceleration. The parallelized version is denoted as Docking_Algorithm$_{\text{GPU}}$:

$$\text{Docking\_Algorithm}_{\text{GPU}}(\text{Input\_Structures}) = \text{GPU\_Parallelize}(\text{Docking\_Algorithm}, \text{Input\_Structures}) \tag{9.54}$$

Here, Input_Structures represent the molecular configurations.

Chemical Formulas

The use of GPU acceleration in chemical formulas:

$$\text{Docking\_Algorithm}_{\text{GPU}}(\text{Input\_Structures}) = \text{GPU\_Parallelize}(\text{Docking\_Algorithm}, \text{Input\_Structures}) \tag{9.55}$$

GPU acceleration significantly reduces the computation time of docking simulations, allowing for the exploration of larger datasets and more complex molecular interactions.

Speedup Calculation

The speedup (Speedup) achieved by GPU acceleration can be calculated using the formula:

$$\text{Speedup} = \frac{\text{Execution Time}_{\text{CPU}}}{\text{Execution Time}_{\text{GPU}}} \tag{9.56}$$

Here, Execution Time$_{\text{CPU}}$ and Execution Time$_{\text{GPU}}$ represent the time taken by the docking algorithm on CPU and GPU, respectively.

The adoption of GPU acceleration in molecular docking not only enhances the speed of simulations but also enables the investigation of larger and more complex biological systems.

9.2.7   Integration with Systems Biology

The trend of integrating molecular docking with systems biology approaches is gaining traction. Researchers aim to simulate and predict ligand-receptor interactions in dynamic cellular environments. This trend aligns with the broader goal of understanding the implications of drug binding within the context of complex biological systems.

9.2.8   Enhanced Sampling Techniques for Conformational Exploration

Conformational exploration is a crucial aspect of molecular docking, allowing the study of various ligand and receptor conformations. Enhanced sampling techniques aim to improve the efficiency of sampling the conformational space, providing a more comprehensive exploration of possible binding modes.

Metadynamics

Metadynamics is a powerful enhanced sampling method that biases the simulation along predefined collective variables (ξ). The metadynamics potential (V_{meta}) is added to the original potential energy (V):

$$V_{\mathrm{total}} = V + V_{\mathrm{meta}}(\xi) \tag{9.57}$$

The metadynamics potential is updated during the simulation, encouraging the exploration of regions not adequately sampled in traditional molecular dynamics.

Chemical Formulas

The representation of metadynamics potential in chemical formulas:

$$V_{\mathrm{total}} = V + V_{\mathrm{meta}}(\xi) \tag{9.58}$$

Metadynamics enhances conformational sampling by flattening the energy landscape along specific collective variables, promoting the exploration of diverse ligand-receptor conformations.

Replica Exchange Molecular Dynamics (REMD)

REMD is another enhanced sampling technique that involves running multiple simulations at different temperatures. Periodic exchanges of conformations between replicas enhance the exploration of conformational space.

$$P_{\text{exchange}} = \min\left(1, e^{-\frac{\Delta E}{k_B T}}\right) \qquad (9.59)$$

Here, ΔE is the energy difference between replicas, k_B is the Boltzmann constant, and T is the temperature.

Chemical Formulas

The representation of replica exchange probability in chemical formulas:

$$P_{\text{exchange}} = \min\left(1, e^{-\frac{\Delta E}{k_B T}}\right) \qquad (9.60)$$

Enhanced sampling techniques significantly improve the exploration of ligand and receptor conformations, providing more accurate insights into the binding process.

9.2.9   3D Printing in Drug Design

Three-dimensional (3D) printing has emerged as a revolutionary technology with applications in various fields, including drug design. The ability to fabricate physical models of molecular structures and drug compounds has opened new possibilities for understanding and optimizing molecular interactions.

Representation of Molecular Structures

In 3D printing, molecular structures can be accurately represented using chemical formulas and molecular graphics. The chemical structure of a drug molecule, represented by its molecular formula $(C_xH_yN_z)$, can be translated into a 3D-printable model.

$$\text{Molecular Formula:} \quad C_xH_yN_z \tag{9.61}$$

The three-dimensional arrangement of atoms within the molecule is crucial for understanding the spatial aspects of drug-receptor interactions.

Chemical Formulas

Representation of a molecular formula for 3D printing:

$$\text{Molecular Formula:} \quad C_xH_yN_z \tag{9.62}$$

3D printing allows researchers and medicinal chemists to physically manipulate and visualize drug molecules, promoting a deeper understanding of their structure-activity relationships.

Customizable Scaffold Designs

3D printing enables the creation of customizable scaffold designs for drug delivery systems. Mathematical formulas describing the geometry of scaffolds can be translated into printable models.

$$\text{Scaffold Geometry:} \quad \text{Mathematical Formula} \tag{9.63}$$

The ability to precisely control the physical characteristics of drug delivery systems enhances the design process.

Mathematical Formulas

Representation of a mathematical formula for scaffold geometry:

$$\text{Scaffold Geometry:} \quad \text{Mathematical Formula} \tag{9.64}$$

Overall, the integration of 3D printing in drug design provides a tangible and interactive approach to studying molecular structures and designing innovative drug delivery systems.

9.2.10 Open Science Initiatives and Collaborative Platforms

The landscape of scientific research is evolving, and open science initiatives play a pivotal role in fostering collaboration, transparency, and accessibility of research findings. Collaborative platforms leverage the power of collective intelligence, enabling researchers to work together seamlessly.

Open Science Principles

Open science is guided by principles that advocate for the sharing of research outputs, methodologies, and data. These principles contribute to the advancement of knowledge by allowing others to scrutinize, reproduce, and build upon existing research.

$$\text{Open Science Principles:} \quad \text{Transparency, Accessibility, Collaboration} \quad (9.65)$$

Researchers adhering to open science principles make their work publicly available, contributing to a more inclusive and collaborative research environment.

Collaborative Platforms

Various collaborative platforms facilitate open science practices, providing spaces for researchers to share ideas, data, and methodologies. These platforms often incorporate version control systems, enabling real-time collaboration and tracking of changes.

$$\text{Collaborative Platforms:} \quad \text{GitHub, GitLab, Overleaf, ResearchGate} \quad (9.66)$$

Integration with version control allows researchers to work collaboratively on manuscripts, code, and datasets, promoting transparency and reproducibility.

Version Control Systems

Version control systems, such as Git, are integral to collaborative platforms. They enable tracking changes, managing different versions of files, and facilitating seamless collaboration among researchers.

$$\text{Version Control Systems:}\quad \text{Git, Mercurial} \tag{9.67}$$

By using version control, researchers can contribute to projects without the risk of conflicting changes, ensuring a streamlined collaborative process.

Mathematical Formulas

Representation of principles and platforms using mathematical notation:

$$\text{Open Science Principles:}\quad \text{Transparency, Accessibility, Collaboration} \tag{9.68}$$

$$\text{Collaborative Platforms:}\quad \text{GitHub, GitLab, Overleaf, ResearchGate} \tag{9.69}$$

$$\text{Version Control Systems:}\quad \text{Git, Mercurial} \tag{9.70}$$

Embracing open science and leveraging collaborative platforms contribute to the democratization of knowledge and the acceleration of scientific progress.

9.2.11 Interpretable AI Models for Decision Support

With the increasing use of artificial intelligence in molecular docking, there is a growing trend towards developing interpretable AI models. Understanding the decisions made by AI algorithms in ligand binding predictions is crucial for gaining insights into the underlying biology. Interpretable models enhance the trustworthiness of AI-driven predictions in drug discovery.

9.2.12 Quantifying Uncertainty in Predictions

In molecular docking, accurately predicting binding affinities is essential for understanding ligand-receptor interactions. However, it is equally important to quantify the uncertainty associated with these predictions. Uncertainty estimation provides insights into the reliability and robustness of computational predictions.

Probabilistic Models

Probabilistic models are instrumental in capturing uncertainty in predictions. Bayesian approaches, for example, allow the modeling of probability distributions over the predicted binding affinities. The uncertainty is then expressed through variance or confidence intervals.

$$\text{Probabilistic Models: Bayesian Regression, Gaussian Processes} \tag{9.71}$$

These models enable the calculation of uncertainty intervals, providing a more comprehensive view of the prediction reliability.

Uncertainty Quantification

Uncertainty quantification involves assigning a level of confidence to predicted values. This can be achieved through various statistical measures, such as standard deviation, confidence intervals, or probabilistic density functions.

$$\text{Uncertainty Measures: Standard Deviation, Confidence Intervals} \tag{9.72}$$

Mathematically, uncertainty quantification is expressed as:

$$\text{Uncertainty} = f(\text{Predicted Affinity}, \text{Model Confidence}) \tag{9.73}$$

Monte Carlo Sampling

Monte Carlo sampling is a widely used technique for uncertainty estimation. By repeatedly sampling the input space and observing variations in predictions, researchers can infer the uncertainty associated with the model.

$$\text{Monte Carlo Sampling:} \quad \text{Sampling from the Predictive Distribution} \quad (9.74)$$

The uncertainty in predictions is then calculated based on the distribution of sampled values.

Mathematical Formulas

Representation of uncertainty quantification using mathematical notation:

$$\text{Probabilistic Models:} \quad \text{Bayesian Regression, Gaussian Processes} \quad (9.75)$$

$$\text{Uncertainty Measures:} \quad \text{Standard Deviation, Confidence Intervals} \quad (9.76)$$

$$\text{Uncertainty} = f(\text{Predicted Affinity}, \text{Model Confidence}) \quad (9.77)$$

$$\text{Monte Carlo Sampling:} \quad \text{Sampling from the Predictive Distribution} \quad (9.78)$$

Quantifying uncertainty in predictions enhances the interpretability of computational results and aids decision-making in drug discovery.

9.2.13   Blockchain Technology for Data Security

In the realm of molecular docking and drug discovery, ensuring the security and integrity of data is paramount. Blockchain technology, originally designed for secure financial transactions, has garnered attention for its potential applications in safeguarding scientific data.

Key Features of Blockchain

Blockchain operates on a decentralized and distributed ledger system. The key features that make it suitable for enhancing data security include:

- **Decentralization:** Data is not stored in a central location but distributed across multiple nodes.

- **Immutability:** Once data is recorded in a block, it cannot be altered, ensuring the integrity of the information.

- **Transparency:** All participants in the network have visibility into the transactions or data entries.

- **Consensus Mechanism:** Changes to the blockchain require agreement among network participants, enhancing security.

Application in Molecular Docking

Implementing blockchain technology in molecular docking scenarios can address concerns related to data integrity and collaboration. Each step in the docking process, from ligand-receptor interactions to scoring functions, can be securely recorded in a blockchain.

$$\text{Blockchain Data Entry:} \quad \text{Block}_i = \text{Hash}(\text{Previous Block}, \text{Data}_i) \quad (9.79)$$

The cryptographic hashing ensures that any attempt to tamper with the data would require altering subsequent blocks, providing a robust defense against unauthorized changes.

Smart Contracts for Data Access

Smart contracts, self-executing contracts with the terms of the agreement directly written into code, can control data access in molecular docking databases. Access permissions and data sharing agreements can be encoded, ensuring only authorized parties can retrieve specific information.

$$\text{Smart Contract:}\quad \text{if(Requesting Party = Authorized) then Grant Access}$$
$$(9.80)$$

By leveraging blockchain and smart contract technologies, the molecular docking community can enhance data security, foster collaboration, and maintain a transparent record of scientific findings.

9.2.14   Advances in Visualization Tools for Molecular Complexes

The visualization of molecular complexes is crucial for gaining insights into the binding interactions between ligands and receptors. Recent advancements in visualization tools have significantly enhanced our ability to explore and understand complex molecular structures.

Mathematical Representation of Molecular Complexes

The mathematical representation of a molecular complex involves the spatial coordinates of atoms, bond lengths, angles, and torsion angles. Given a molecular complex C with atoms $A_1, A_2, \ldots, A_n$, the spatial coordinates can be represented as:

$$C = \{(x_1, y_1, z_1), (x_2, y_2, z_2), \ldots, (x_n, y_n, z_n)\} \qquad (9.81)$$

This representation forms the basis for visualization algorithms.

Differential Evolution for Molecular Visualization

Differential Evolution (DE) is a powerful optimization algorithm that has found applications in molecular visualization. DE can be applied to refine the orientation and position of ligands within a binding site for improved visualization.

$$\text{DE Algorithm:}\quad \vec{v}_{i,G+1} = \vec{x}_{r1,G} + F \cdot (\vec{x}_{r2,G} - \vec{x}_{r3,G}) \qquad (9.82)$$

Here, $\vec{v}_{i,G+1}$ represents the trial vector for the ligand, $\vec{x}_{r1,G}$, $\vec{x}_{r2,G}$, and $\vec{x}_{r3,G}$ are randomly selected vectors, and F is the differential weight.

Chemical Formulas

The visualization of molecular complexes in chemical formulas:

$$\text{Ligand} + \text{Receptor} \xrightarrow{\text{DE}} \text{Optimized Ligand} \tag{9.83}$$

This equation illustrates the application of the DE algorithm in optimizing ligand conformation within the receptor binding site.

Interactive 3D Visualization Tools

Advancements in interactive 3D visualization tools, such as PyMOL, Chimera, and VMD, enable researchers to manipulate and explore molecular complexes in real-time. These tools provide a user-friendly interface for visualizing complex interactions.

Immersive Virtual Reality (VR) Environments

Immersive VR environments offer a transformative way to visualize molecular complexes. Researchers can navigate through complex structures in a virtual space, providing a more intuitive and immersive experience.

$$\text{VR Visualization:} \quad \text{Immersive Experience} \to \text{Enhanced Structural Understanding} \tag{9.84}$$

The integration of mathematical algorithms, DE optimization, and interactive visualization tools has ushered in a new era of molecular complex exploration, offering unprecedented insights into drug-receptor interactions.

9.2.15 Multi-Target and Polypharmacology Considerations

In drug discovery, considering the interactions of a compound with multiple targets, known as polypharmacology, has gained increasing attention. Multi-target approaches aim to address the complexity of biological systems by designing compounds that can modulate multiple targets simultaneously.

Polypharmacology in Drug Design

Polypharmacology involves designing drugs that interact with more than one target to achieve enhanced therapeutic effects or address multiple disease pathways. Mathematically, polypharmacology can be represented as:

$$\text{Polypharmacology:} \quad \text{Compound} \rightleftharpoons \text{Target}_1 + \text{Target}_2 + \ldots {}^+ \text{Target}_n \quad (9.85)$$

This representation illustrates the interaction of a compound with multiple targets.

Mathematical Optimization for Multi-Target Design

Mathematical optimization techniques play a crucial role in designing compounds with desired interactions across multiple targets. An optimization problem for multi-target drug design can be formulated as:

$$
\begin{aligned}
\underset{\mathbf{x}}{\text{minimize}} \quad & f(\mathbf{x}) \\
\text{subject to} \quad & g_i(\mathbf{x}) \leq 0, \quad i = 1, 2, \ldots, m,
\end{aligned}
\quad (9.86)
$$

where $\mathbf{x}$ represents the molecular structure parameters, $f(\mathbf{x})$ is the objective function representing the balance between affinities to different targets, and $g_i(\mathbf{x})$ are constraints ensuring drug-like properties.

Chemical Formulas

The polypharmacological interaction of a compound with multiple targets in chemical formulas:

$$\text{Compound} + \text{Target}_1 \rightleftharpoons \text{Complex}_1, \quad \text{Compound} + \text{Target}_2 \rightleftharpoons \text{Complex}_2, \ldots$$
$$(9.87)$$

These equations show the binding of the compound to different targets, forming distinct complexes.

Network Pharmacology

Network pharmacology provides a systematic approach to understanding the relationships between drugs, targets, and diseases in a network framework. It considers the holistic impact of drugs on the entire biological system.

$$\text{Network Pharmacology:} \quad \text{Integration of Drug-Target-Disease Interactions} \tag{9.88}$$

The integration of multi-target and polypharmacology considerations in drug design holds promise for developing more effective and versatile therapeutic agents.

9.2.16 Ethical Considerations in AI-Driven Drug Discovery

The integration of artificial intelligence (AI) in drug discovery brings forth ethical considerations that require careful examination. As AI technologies advance, it becomes essential to address potential ethical challenges associated with their application in pharmaceutical research.

Patient Privacy and Data Security

The utilization of large-scale datasets, including patient information, raises concerns about privacy and data security. Ensuring the confidentiality of patient data is paramount. Encryption, anonymization, and strict access controls are critical measures to safeguard sensitive information.

Transparency and Explainability

AI algorithms often operate as complex black-box models, making it challenging to interpret their decision-making processes. Ensuring transparency and explainability in AI-driven drug discovery is crucial for building trust among researchers, clinicians, and regulatory bodies. It involves providing clear insights into how AI models arrive at specific predictions or recommendations.

Bias and Fairness

The datasets used to train AI models may inadvertently contain biases, leading to biased predictions. Addressing issues related to bias and fairness is essential to prevent disparities in drug discovery outcomes. Regular audits and evaluations of AI algorithms for potential biases are necessary.

Informed Consent

In cases where AI technologies involve the use of patient data, obtaining informed consent becomes a pivotal ethical consideration. Researchers must ensure that patients are adequately informed about the purposes and potential implications of utilizing their data in AI-driven drug discovery.

Responsible AI Research and Development

Promoting responsible AI research and development involves adhering to ethical guidelines, regulatory standards, and industry best practices. Researchers and developers should prioritize the ethical implications of their work and actively engage in discussions around the responsible use of AI in drug discovery.

Chemical Formulas

Representing the ethical considerations in chemical formulas:

$$\text{Ethical Considerations:} \quad \text{AI+Drug Discovery} \rightleftharpoons \text{Responsible Research and Development} \tag{9.89}$$

These equations symbolize the integration of AI into drug discovery while emphasizing the importance of ethical principles.

Collaboration and Open Dialogue

Fostering collaboration and open dialogue among stakeholders, including researchers, ethicists, policymakers, and the public, is vital. Establishing frameworks for continuous communication ensures that ethical considerations evolve alongside advancements in AI-driven drug discovery.

Ethical considerations play a central role in shaping the future of AI applications in drug discovery, emphasizing the need for a balanced and responsible approach.

Chapter 10

Appendix

10.1 Glossary

This glossary provides a comprehensive reference for key terms and concepts used in the field of molecular docking. It serves as a quick guide for readers to enhance their understanding of the terminology employed throughout the book.

- **Molecular Docking:** A computational method used to predict the preferred orientation and binding affinity of a ligand to a target receptor.

- **Scoring Function:** A mathematical algorithm that evaluates and ranks different ligand poses based on their likelihood of binding to a target receptor.

- **Ligand:** A small molecule that binds to a biological macromolecule, such as a protein, to form a complex.

- **Receptor:** A biomolecule, often a protein, that interacts with a ligand through specific binding sites.

- **Binding Affinity:** The strength of the interaction between a ligand and a receptor, indicating how likely they are to form a stable complex.

- **Conformation:** The three-dimensional arrangement of atoms in a molecule, representing its specific shape.

- **Grid-Based Methods:** Molecular docking approaches that discretize space into a grid to explore possible ligand binding orientations.

- **Flexible Docking:** Docking studies that account for flexibility in both ligands and receptors, allowing for a more realistic simulation.

- **Force Field:** A set of parameters and equations used to describe the energy of a molecular system, guiding simulations in molecular dynamics.

- **Hydrogen Bond:** A type of non-covalent interaction between a hydrogen atom attached to an electronegative atom and another electronegative atom.

- **Pose:** A specific conformation or orientation adopted by a ligand when bound to a receptor.

- **Virtual Screening:** The computational screening of large compound libraries to identify potential drug candidates based on their predicted binding affinities.

- **Bioinformatics:** The application of computational techniques to analyze and interpret biological data, often applied in molecular docking studies.

- **Quantum Mechanical Docking:** Docking studies that incorporate quantum mechanics principles for a more accurate representation of electronic interactions.

- **Fragment-Based Drug Design:** An approach in drug discovery that starts with small molecular fragments, which are then optimized into lead compounds.

- **GPU Acceleration:** The use of Graphics Processing Units (GPUs) to accelerate computational calculations, improving the speed of docking simulations.

- **Deep Learning:** A subset of machine learning that involves training neural networks on large datasets to make predictions, increasingly applied in scoring functions.

- **Systems Biology:** The interdisciplinary study of complex biological systems, integrating computational and experimental approaches.

- **3D Printing in Drug Design:** The utilization of 3D printing technology to create personalized drug delivery systems guided by molecular docking studies.

- **Blockchain Technology:** A decentralized and secure method for storing and sharing data, gaining traction in ensuring data integrity in molecular docking research.

This glossary serves as a valuable resource for readers navigating the terminology associated with molecular docking, ensuring clarity and comprehension throughout their exploration of the handbook.

10.2 Abbreviations

This section provides a list of abbreviations used throughout the book, aiding readers in quickly understanding and referencing key terms.

- **MD**: Molecular Docking - A computational technique to predict the binding of molecules.

- **SF**: Scoring Function - An algorithm to evaluate the fitness of a ligand-receptor complex.

- **L**: Ligand - A molecule that binds to a specific site on a receptor.

- **R**: Receptor - A macromolecule, often a protein, that interacts with a ligand.

- **BA**: Binding Affinity - The strength of the interaction between a ligand and a receptor.

- **Conf**: Conformation - The spatial arrangement of atoms in a molecule.

- **GBM**: Grid-Based Methods - Docking approaches that discretize space for efficient exploration.

- **FD**: Flexible Docking - Considering the flexibility of ligands and receptors in docking studies.

- **FF**: Force Field - A mathematical model used to describe the potential energy of a molecular system.

- **HB**: Hydrogen Bond - A type of chemical bond critical in molecular recognition.

- **P**: Pose - A specific orientation of a ligand in the binding site.

- **VS**: Virtual Screening - Using computational methods to screen large compound libraries.

- **Bioinfo**: Bioinformatics - The application of computational techniques to biological data.

- **QMD**: Quantum Mechanical Docking - Incorporating quantum mechanics principles in docking simulations.

- **FBDD**: Fragment-Based Drug Design - Designing drugs based on small molecular fragments.

- **GPU**: Graphics Processing Unit - Accelerating computational calculations in docking studies.

- **DL**: Deep Learning - Using neural networks to improve scoring functions.

- **SB**: Systems Biology - Studying complex biological systems using computational and experimental methods.

- **3DPDD**: 3D Printing in Drug Design - Utilizing 3D printing for personalized drug delivery systems.

- **BT**: Blockchain Technology - Ensuring data integrity in molecular docking research through decentralized secure methods.

This list will help readers navigate the book with ease, enhancing their understanding of the material.

10.3 References

This section provides a comprehensive list of references cited throughout the book, aiding readers in further exploring the topics covered. The references include seminal works, research papers, and relevant literature in the field of molecular docking.

1. Jones, G., Willett, P., Glen, R. C., Leach, A. R., & Taylor, R. (1997). Development and validation of a genetic algorithm for flexible docking. Journal of Molecular Biology, 267(3), 727-748.

2. Morris, G. M., Huey, R., Lindstrom, W., Sanner, M. F., Belew, R. K., & Goodsell, D. S. (2009). AutoDock4 and AutoDockTools4: Automated docking with selective receptor flexibility. Journal of Computational Chemistry, 30(16), 2785-2791.

3. Trott, O., & Olson, A. J. (2010). AutoDock Vina: Improving the speed and accuracy of docking with a new scoring function, efficient optimization, and multithreading. Journal of Computational Chemistry, 31(2), 455-461.

4. Wang, R., Lu, Y., & Wang, S. (2003). Comparative evaluation of 11 scoring functions for molecular docking. Journal of Medicinal Chemistry, 46(12), 2287-2303.

5. Leach, A. R., & Gillet, V. J. (2007). An introduction to chemoinformatics (Vol. 38). Springer.

6. Kitchen, D. B., Decornez, H., Furr, J. R., & Bajorath, J. (2004). Docking and scoring in virtual screening for drug discovery: methods and applications. Nature Reviews Drug Discovery, 3(11), 935-949.

7. Wang, Z., Sun, H., Yao, X., Li, D., & Xu, L. (2021). Computationally Simulating Protein–Ligand Interactions: A Review. Computational and Structural Biotechnology Journal, 19, 2541-2553.

8. Dixon, S. L., Smondyrev, A. M., Knoll, E. H., Rao, S. N., Shaw, D. E., & Friesner, R. A. (2006). PHASE: a new engine for pharmacophore perception, 3D QSAR model development, and 3D database screening: 1. Methodology and preliminary results. Journal of Computer-Aided Molecular Design, 20(10-11), 647-671.

9. Sterling, T., & Irwin, J. J. (2015). ZINC 15–Ligand Discovery for Everyone. Journal of Chemical Information and Modeling, 55(11), 2324-2337.

10. Salentin, S., Schreiber, S., Haupt, V. J., Adasme, M. F., & Schroeder, M. (2015). PLIP: fully automated protein–ligand interaction profiler. Nucleic Acids Research, 43(W1), W443-W447.

The references encompass various aspects of molecular docking, from algorithm development to practical applications. Readers are encouraged to explore these works for a deeper understanding of the subject matter.